EARLY ASTRONOMY:
FROM BABYLONIA
TO COPERNICUS

Abd al-Karim's Syrian Astrolabe.   A large thirteen
inch astrolabe of brass inlaid with silver and copper
made by Abd al-Karim al-Misri al-Usturlabi in
AD 1235 for a Sultan in Damascus. A peculiar feature is
the use of animal and human figures for many of the
star-pointers. (Reproduced by permission of the
Trustees of the British Museum where the instrument
is now located.)

EARLY ASTRONOMY

FROM BABYLONIA
TO COPERNICUS

W. M. O'NEIL

SYDNEY UNIVERSITY PRESS

SYDNEY UNIVERSITY PRESS
Press Building, University of Sydney
UNITED KINGDOM, EUROPE, MIDDLE EAST, AFRICA
HB Sales, Enterprise House, Ashford Road, Ashford
Middlesex TW15 1XB, England
NORTH AND SOUTH AMERICA
International Specialized Book Services, Inc.
5602 N.E. Hassalo Street, Portland
OR 97213-3640, United States of America

National Library of Australia Cataloguing-in-Publication data

O'Neil, W. M. (William Matthew), 1912–
Early astronomy: from Babylonia to Copernicus.

Bibliography.
Includes index.
ISBN 0 424 00117 9.

1. Astronomy – History. I. Title.

520'.9

First published 1986
© W. M. O'Neil 1986
Printed in Australia at Griffin Press Limited
Marion Road, Netley, South Australia

CONTENTS

List of Figures   vii

List of Tables   ix

Preface   xi

1   Introduction   1

2   Early Babylonian Astronomy   15

3   Late Babylonian Astronomy   35

4   Early Greek Astronomy   52

5   Hipparchos   67

6   Ptolemy   82

7   The Aftermath of Hellenistic Astronomy   99

8   Astronomy and Astrology   109

9   Arabian Astronomy and its Importance
for the Revival in the West   117

10   Early Astronomical Instruments   133

11   The Reawakening of the West   146

12   Constellation and Star Names from
Ancient to Modern Times   153

13   Copernicus   164

Appendix 1   Chinese Astronomy   176

Appendix 2   Possible Megalithic Observatories   181

Glossary   197

Bibliography   203

Index   210

FIGURES

1.1 The apparent paths of four stars for an observer at 30°N   2
1.2 The celestial sphere, the celestial equator and the ecliptic   4
1.3 The variation of the Moon's apparent path in latitude   5
1.4 The direct and retrograde phases of Mars   6
1.5 Periods of invisibility of a superior and an inferior planet   7
2.1 The main stars in our *Taurus* and in the Babylonian *Bull-of-Heaven*   18
2.2 The early Babylonian 36 stars   20
2.3 The periods of visibility and invisibility of Venus   22
2.4 The basis of the resonance period of Venus   28
2.5 The shadow of the Earth involved in lunar eclipses   32
3.1 Babylonian numerals   38
3.2 Graph of the first stations of Jupiter illustrating a step function   45
3.3 Graph of the second stations of Jupiter illustrating a zigzag function   47
4.1 Aristarchos's method of measuring the distance of the Sun in terms of the Moon's distance   58
4.2 Eratosthenes's method of measuring the circumference of the Earth   59
4.3 Eudoxos's homocentric spheres for the Sun and the Moon   62
4.4 Eudoxos's homocentric spheres for Jupiter   63
4.5 (A) The hippopede produced by the two inner spheres. (B) The hippopede alternately stretched and compressed as a result of the second sphere   64
5.1 The westward shift of the equinoxes along the ecliptic over time   71

5.2 Figures produced by a point on the epicycle which is rotating in the same or the opposite direction to the deferent and at various relative rates 77

5.3 Hipparchos's explanation of the anomaly of the Sun 78

5.4 Hipparchos's attempted explanation of the anomaly of the Moon 79

6.1 Ptolemy's theory of the Moon 86

6.2 The major general features of Ptolemy's theories of the planets other than the Moon 88

6.3 The detailed features of Ptolemy's theory of Mars 89

6.4 An epicycle and a deferent by modern reckoning rotating in the same direction and the same period; by ancient reckoning the epicycle would have been deemed not to have rotated at all 90

6.5 The relative positions of the centre of the deferent, of the Earth and of the equant in the main Ptolemaic planetary theories 90

6.6 Ptolemy's theory of Mercury 92

9.1 The 'Tusi couple' as used by al-Shatir in his theory of Mercury 129

10.1 An early Egyptian shadow–clock 135

10.2 A constant head clepsydra 137

10.3 The triquetrum or *regula Ptolemaica* 141

10.4 An armillary sphere 142

10.5 The components of a typical Arabian planispheric astrolabe 143

12.1 The central figure or *imago* of *Orion* 158

13.1 Copernicus's explanation of the retrograde phase of a planet 165

13.2 In Copernicus's heliofocal theory inferior planets never stray far ahead of or behind the Sun 166

13.3 Copernicus's analysis of the orbit of the Earth 168

13.4 Copernicus's analysis of the orbit of Mars 168

13.5 Copernicus's theory of the Moon 171

13.6 Kepler's first law 175

13.7 Kepler's second law 175

A2.1 For an observer at 60°N the Sun will rise at the summer solstice 46°54′ north of east 182

A2.2 The effect of the height of the horizon on the point of sunrise for an observer away from the equator 183

A2.3 The alignment claimed by Thom for a winter solstice sunset at Kintraw 187

TABLES

2.1   The 31 Babylonian reference stars   26
2.2   Part of a Babylonian fragment recording possible lunar eclipses   30–1
2.3   Dates and types of lunar eclipses from 1940 to 1959   33
3.1   Part of a table giving first stations of Jupiter   44
3.2   Dates and positions of second stations of Jupiter   46
3.3   Part of a late Babylonian luni-solar ephemeris   48
5.1   Frequencies recorded in Ptolemy's catalogue for stellar longitudes and latitudes   75
6.1   The ratios in Ptolemy's theories of the radius of the epicycle to that of the deferent   91
6.2   Ptolemy's estimates of the minimum and maximum planetary distances   98
9.1   Relative lengths of the radii in al-Shatir's planetary theories   131
12.1   Names of zodiacal constellations in several languages   154–5
12.2   Names of the seven planets in several languages   156
13.1   Copernicus's values for the mean distances of the planets from the Sun   167
13.2   The ratios of the radii of the epicycles and of the deferent according to al-Shatir and Copernicus   170

PREFACE

THIS SMALL BOOK is aimed primarily at the interested general reader. It may also be of some use to students beginning the study of the early history of science. It deals with the development of astronomy within the tradition begun in Babylonia, transformed by the Greeks, preserved by the Arabians and brought to an end by Copernicus who opened the door for modern astronomy.

Parts of the story have been told in some detail in many works. Dreyer (1905) is very good on the Greeks, informative on the Arabians and very good on Copernicus; he is silent on the Babylonians. Neugebauer (1969, 1975) is good on the Babylonians and excellent on the Greeks. Pannekoek (1961) is good on the Babylonians, on the Greeks and on Copernicus and his immediate predecessors but is sketchy on the Arabians; much more material has become available on the Babylonians and on the Arabians since he wrote his original Dutch version published in 1951. There are numerous excellent works devoted to the Greeks such as those by Heath (1913), Dicks (1970) and Lloyd (1970, 1973). The *Dictionary of Scientific Biography* is a mine of information about individual Greek, Arabian and late medieval European astronomers.

While there is a good supply of secondary materials, the supply of primary materials is patchy. We have the treatises by Ptolemy and by Copernicus, both now available in English translations. Only one work by Hipparchos, Ptolemy's great predecessor, has been preserved; it is a minor work and is available in German translation. A minor pre-Ptolemic treatise by Geminos has been preserved and is available in German translation. Commentaries on Ptolemy's *Syntaxis mathematike* are available in French translations by Rome. Many Babylonian 'records' have been found.

These are records of observations or records of predictions. There is little in the way of accounts of Babylonian techniques and methods and no general treatise. We know little, possibly nothing, of individual Babylonian astronomers. An increasing number of Arabian texts have become available in English translation in recent years.

As may be inferred from this cursory survey of the materials, I have had to rely heavily on secondary sources. Even where I have turned to a primary source, I have had to resort to a translation into a language I can read – preferably English, but in its absence French, German and Latin where I have only halting skills. I rely entirely on others for what I report from untranslated Greek, Sanskrit, Arabic and Chinese sources.

Though I am writing in the southern hemisphere, the sky-watchers about whom I write lived in the northern hemisphere. So I describe celestial events from a northern viewpoint.

I am grateful to several readers of the draft material in several stages of production. Dr Peter Wilson, a psychologist with an amateur interest in astronomy, and Dr Bruce McAdam, a radio astronomer, read an early version and suggested many alterations in the interests of improved clarity. Dr McAdam suggested a substantial rearrangement of material which resulted in the shifting of what had been an early chapter on the megalithic monuments in Britain and Brittany to an appendix, as many technical matters relevant to that discussion had been more gently introduced earlier. I submitted, prematurely I now recognize, a draft for general assessment by Sydney University Press. One reader of this draft wrote a long and generally unfavourable report. He drew attention to a number of careless errors which I should not have let slip through. He maintained that I had given too single-minded an account of R. R. Newton's attacks on Ptolemy; I have now toned down my support of Newton's case. I have changed some of what he regarded as dubious usages, once or twice with reluctance. I have rejected some of his proposals either because he rather than I was in error or because I did not accept his view as to how history should be written.

A reader of the next revised version who was quite encouraging suggested several additions of material, two of which I have elected not to adopt but several of which I have included. Perhaps the most important of these additions is a glossary of technical terms. When I had introduced a technical term I gave a definition of it, but just used it on a second or later occasion. It was suggested that the reader who had forgotten the meaning and where to turn back for its elucidation would be helped by a glossary.

I have a special debt to Fisher Library in the University of Sydney. Virtually all the material I needed was available in Fisher.

The manuscript has been written many times, and typed many times in part or in full. I have to thank Miss A. D. O'Connor, Mrs Sheina Brunsman, Mrs Maureen Farago, Miss Eileen Hunt and Mrs Cathy Tonner for contributing to this typing marathon.

Four of the numerous figures have been reproduced with the permission of Oxford University Press, a few others have been reproduced from other publications of mine, but the great majority of them have been drawn from my rough sketches by Mr John Roberts, to whom I am very grateful.

Once again I have to express gratitude to my wife, first for putting up with my obsession with astronomical matters which often led to my withdrawal from domestic responsibilities, and second for her reading of the near final draft with an eye to expression, usage and spelling.

W. M. O'Neil
The University of Sydney
January, 1986

INTRODUCTION

Most, perhaps all, early peoples have taken an interest in events in the sky. They have identified and named the brighter celestial bodies and sets (constellations) of them. They have noted their apparent motions, usually distinguishing between the fixed stars and the wanderers. Quite early, possibly earlier than 4000 BC in Mesopotamia, it was noticed that most of the stars crossed the night sky in apparently fixed formation though rising and setting a little earlier than the Sun day by day. It was also noticed quite early that stars towards the north never rose across the horizon or set below it. Instead they were seen to trace during the night arcs of circles which never dipped below the horizon; we call such stars circumpolar. The proportion of fixed stars that rise or set depends on the terrestrial latitude of the observer. For an observer at the terrestrial equator, there are virtually no circumpolars (atmospheric refraction may lift a few stars above the horizon); for an observer 30°N, the north celestial pole is 30° above the horizon and the circumpolars occupy a circle with a radius of 30° around it; for an observer 45°N these values increase to 45°.

The fixed stars other than the circumpolars rise and set some four minutes in modern terms per day progressively ahead of sunrise and sunset. Consequently after their last appearance when setting after sunset (their heliacal setting) they become invisible (blotted out when they next rise by the brighter Sun) until their next first appearance before sunrise (their heliacal rising). The duration of their invisibility depends on their position north or south of the celestial equator and the observer's latitude north or

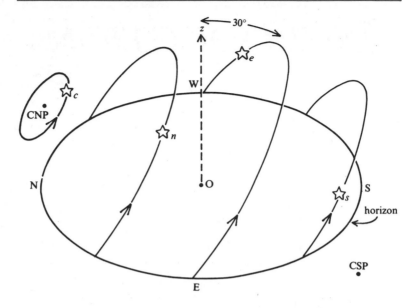

Fig. 1.1 *The apparent paths of four stars for an observer, 0, at 30°N in terrestrial latitude. The zenith, z, is directly over the observer's head. A star such as e which rises due east and sets due west traces the celestial equator which is tilted 30°S of the zenith; its visible path is half a circle. Stars such as n and s also follow a path tilted south by 30°, n tracing more than half a circle and s less than half above the horizon. A star such as c, a circumpolar, never cuts the horizon; the centre of its circle is the Celestial North Pole (CNP). Below the southern horizon is the Celestial South Pole (CSP).*

south of the terrestrial equator. Stars well to the south in the sky have a longer period of invisibility for an observer in the northern hemisphere than stars more northern in the sky; this effect is increased the more northern the latitude of the observer.

The stars other than the circumpolars each rise at a fixed point on the eastern horizon and set at a corresponding point on the western horizon. These points are determined by the star's declination in the sky (its position north-south of the celestial equator), which slowly changes over time, and the observer's latitude on the Earth (his position north-south of the terrestrial equator). A star on the celestial equator rises due east and sets due west except for an observer near the terrestrial poles. A star $23\frac{1}{2}°$ north of the celestial equator will rise $23\frac{1}{2}°$ north of east and set $23\frac{1}{2}°$ north of west for an observer at the terrestrial equator, but will rise further north of east and set further north of west for an observer north of the terrestrial equator. Such a star will rise over 27° north of east for an observer at 30°. For observers north of the terrestrial equator, the fixed stars trace arcs of circles tilted back from the vertical by an angle equal to the observer's terrestrial latitude (see Fig. 1.1).

In addition to the Sun's apparent motion eastward amongst the fixed stars (by an average of almost 1° per day), the Sun also appears to move north and south of the celestial equator by about $23\frac{1}{2}°$ in half a year. In the northern hemisphere spring the Sun crosses the equator going north. In autumn it crosses the equator going south. These passages mark the spring and autumn equinoxes respectively. On these occasions daylight (from sunrise to sunset) equals night (the two twilights and the period of darkness when the stars are shining). Also on these occasions the Sun rises due east and sets due west and so traces the celestial equator. Approximately midway in times between the equinoxes the Sun reaches its furthest point north of the equator and its furthest point south. These are the summer and the winter solstices respectively; the Sun pauses or stands for a day or two in its northern or southern journey before turning (see Fig. 1.2).

Whereas the Sun takes almost 365.2564 days to make a circuit from one fixed star and back again, it takes almost 365.2422 days to move from one solstice or equinox and back again. The former is the sidereal year and is very slowly increasing by 0.00000012 days per century; the latter is the tropical year (from the Greek *trope*, to turn) and is slowly decreasing by 0.00000614 days per century. The difference in the duration of these two years results from the fact that the equinoctial crossing points are sliding backwards (westward) along the ecliptic by 50″.2 per annum or by 360° in a little over 25 800 years.

The Sun's apparent path amongst the fixed stars is called the ecliptic because eclipses occur only when the Moon is near this plane. The ecliptic is offset from the celestial equator by 23°27′, an angle referred to as the obliquity of the ecliptic. This angle is decreasing very slowly by 0″.47 per annum.

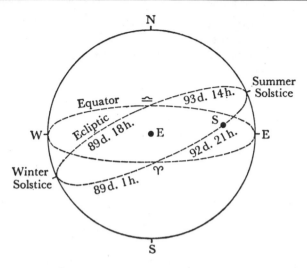

Fig. 1.2 *The continuous circle represents the apparent celestial sphere. The ellipses represent the celestial equator and the ecliptic (the apparent path of the Sun). The positions on the ecliptic of the winter solstice, the summer solstice, the spring equinox (the ram's horns symbol) and the autumn equinox (the balance symbol) are marked. The equinoxes slide westward (anticlockwise) along the ecliptic by 1° in a little less than 72 years.*

Though the Sun moves eastward amongst the fixed stars by an average of almost 1° per day, it advances at a slightly variable rate. Today the Sun spends 92.72 days between the spring equinox and the summer solstice, 93.66 days between the summer solstice and the autumn equinox, 89.84 days between the autumn equinox and the winter solstice, and 88.98 days between the winter solstice and the spring equinox. These differences seem to have been detected in rough terms by the Greeks in the fifth century BC. By the second century BC Hipparchos had these seasonal durations with much greater accuracy. We now know that they change over time as a result of the precession of the equinoxes discovered by Hipparchos and of the motion of the Earth's perihelion (discovered by the Arabs but thought by them to be motion of the perigee of the Sun). The period from one perihelion to the next is 365.25964 days, the anomalistic year.

4

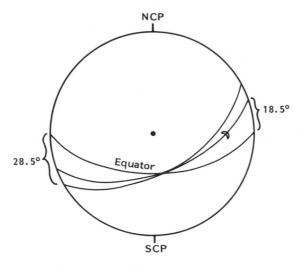

Fig. 1.3 *The variation of the Moon's apparent path between 18°.5 and 28°.5 from the celestial equator, that is between about ± 5° from the ecliptic.*

The Moon also appears to wander eastwards amongst the fixed stars in a period which varies markedly around a mean value of 27.322 days (the sidereal month). The Moon catches up with the Sun in a period averaging a little over 29.53 days and varying from 29.26 to 29.80 days (the synodic month, so called from the Greek word *sunodos*, a meeting). The Moon also wanders north and south, deviating from the Sun's apparent path, the ecliptic, by up to a little over ± 5°; hence the Moon's maximum deviation north or south of the celestial equator may be as much as 28°36' or as little as 18°18', that is by 23°27' ± 5°9' (see Fig. 1.3). The pattern of the Moon's apparent motion amongst the fixed stars is much more variable and complex than that of the Sun and presents greater difficulties for analysis and explication as we shall see in later chapters.

Five other star-like wanderers were recognized quite early: Mercury, Venus, Mars, Jupiter and Saturn, to use their modern names based on Latin. Their wanderings are even more complex than that of the Moon. They appear to deviate, as seen from the Earth, by various amounts from the ecliptic, Venus having the greatest deviation, some ± 8°30'. Whereas the Sun and the Moon appear always to move eastward (though with variable velocities)

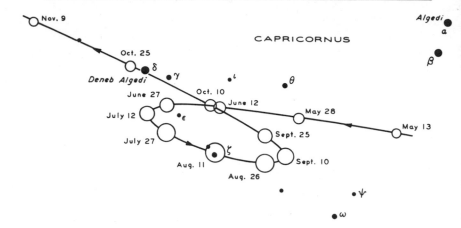

Fig. 1.4   *The direct (west to east) and retrograde (east to west)
phases of Mars over a seven-month period in 1971. The open
circles represent Mars and their size indicates its apparent
brightness. The filled circles are fixed stars. Mars was at its first
station on 12 July, in opposition to the Sun on 11 August
(when it occulted zeta* Capricornii) *and at its second station
on 10 September.*

amongst the fixed stars, the five star-like planets appear to go
through the following phases: for the greater part of the time an
eastward apparent motion (the direct phase) at first increasing and
then decreasing in apparent velocity; a first stationary point; a short
westward apparent motion (the retrograde phase) with an at first
increasing and then decreasing velocity; a second stationary point;
and a resumption of the eastward apparent motion (the direct
phase) (see Fig. 1.4).

Each of these five planets becomes invisible sometime before
until sometime after meeting with the Sun (conjunction). Such a
conjunction occurs twice in each circuit of the stars for Mercury
and Venus, once when each is between the Earth and the Sun
(inferior conjunction) and once when each is on the other side of
the Sun from the Earth (superior conjunction). On the former
occasion the period of invisibility is relatively short and on the
latter much longer (see Fig. 1.5). The other three have only one
period of invisibility, namely when they are in conjunction on the

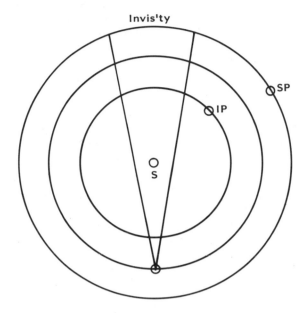

Fig. 1.5   *A superior planet (SP) has a period of invisibility from the Earth when it is near to the line of sight to the Sun; an inferior planet (IP) has two periods of invisibility, one a short period when between the Earth and the Sun (inferior conjunction) and the other a much longer period when on the other side of the Sun (superior conjunction).*

other side of the Sun (they never pass between the Earth and the Sun).

Each of the planets has its own characteristic average period for its circuit of the fixed stars (its sidereal period). For Mercury and Venus this period as seen from the Earth is one year, for Mars about 1.88 years, for Jupiter about 11.86 years and for Saturn about 29.46 years; some of them, for example Mercury and Mars, have greater variations around these average periods than do others. Mercury and Venus as seen from the Sun have mean sidereal periods of 87.97 days and 224.70 respectively, and are seen from the earth to catch up with the Sun in mean periods of 115.88 days and 583.92 days respectively (their synodic periods), whereas the other three, moving with less apparent velocity than the Sun, are caught up by it in 779.94 days in the case of Mars, 398.88 days in

7

the case of Jupiter and 378.09 in the case of Saturn. The retrograde phases of Mercury and Venus occur around the inferior conjunctions of these planets whereas for the other three they occur around opposition.

The periods of invisibility for Mercury and for Venus begin after the planet is last seen rising before sunrise and ends when it is first seen setting after sunset. Each of the other three planets begins its invisibility after it is last seen setting after sunset and ends when it is first seen rising before sunrise (its heliacal rising).

The seven wanderers move through the fixed stars in a band ± 8.5° from the ecliptic which is, as has been reported, now at an angle of about 23°27′ to the celestial equator. This band is called the zodiac, following the Greek *kyklos zodiakos*, the circle of the beasts: the ram, the bull, the twins, the crab, the lion and so on. The Greeks, however, did not invent this notion; it is Mesopotamian as are the names of most if not of all the so-called 'beasts' (only one is not animate, *Libra*, three are human, *Gemini*, *Virgo* and *Aquarius*, and one is part human, *Sagittarius*).

The concern of early peoples with the celestial bodies, their apparent motions and the formations of fixed stars (constellations) may have had many motivations. Curiosity about the spectacle presented in the sky may have been one of them. There would seem also to be several quite common practical motives.

One is concerned with time-reckoning (see Nilsson 1920 and O'Neil 1975). The rising or the setting of the Sun marks the beginning of a day, defined as a cycle of daylight and night darkness. The amount of sky traversed by the Sun between sunrise and sunset provides an indication of how much of the sunlight period has passed. A rough judgement can be made by the unaided eye but something more precise is enabled by a shadow-stick or gnomon casting a shadow on some graduated 'meter' (that is, one marked off in steps), either plane or curved. A sighting device such as a quadrant or sextant may also be used. The latter has the advantage of enabling the observation of angles above the horizon of stars at night as well as angles above the horizon of the Sun by day.

The Egyptians in, say, 2000 BC (see Neugebauer 1969 and Parker 1974), hit on an interesting scheme for defining the hours of the night. They selected a series of stars, probably south of the ecliptic, which rose (and set) at approximately 40-minute intervals

8

(in our time scheme). After ten days each of these stars rose about 40 minutes earlier. The rising of star *A* marked the end of the first night hour for ten days, thereafter it was replaced by its successor *B*, rising 40 minutes earlier in relation to sunset. Thus the 'striking' of the hour by the designated star progressively advanced by about 40 minutes in the course of ten days. This series of stars, of which we can be certain of only a few, were called by the Greeks *dekanoi* (decans) because they were replaced successively at ten-day intervals.

The Egyptians had a strong interest in the heavens but never developed a scientific astronomy. In addition to the decans series, they used the first visible rising of Sirius (called by them Sopdet) as a herald of the rise of the Nile and a beginning of a year of the seasons. The interval between successive heliacal risings of Sirius at Memphis is only slightly greater than the tropical year (365.257 days as against 365.2424 days in 1500 BC), so that the heliacal rising of Sirius occurred about 25 days after the summer solstice, an interval that increased by about a day in 120 years or about five days in six centuries. This is a much poorer equation than the Babylonian and later Greek equation of nineteen years with 235 synodic months; it was, of course, earlier and hence may be expected to be inferior.

The Egyptians identified and named many constellations, usually quite differently from the Babylonian conventions. For example the Big Dipper or *Ursa Major* was identified and named for the hind leg of an ox; other recognized constellations were identified with and named for such Egyptian animals as the alligator and the hippopotamus. Egyptian astronomy made little progress beyond its pre-scientific stage, possibly because of the primitive state of Egyptian mathematics which was effective enough with practical computational problems but of little value with the abstract analyses made by Babylonian and Greek astronomy.

The phases of the Moon provided the Babylonians with a time unit larger than the day but shorter than the month. The intervals between first visible crescent Moon after sunset, the first half Moon, the full Moon, the second half Moon and the last visible crescent before sunrise are on average about seven days (a week) apart. Today the term 'New Moon' refers to the Moon when it is in conjunction with the Sun and consequently invisible. The ancients considered the waxing new Moon to be the first visible crescent just after sunset, and the waning old Moon to be the last visible crescent just before sunrise; there would ordinarily be an interval of about two or three days with no visible Moon. These roughly seven-day quarters of the Moon may well be the basis for our seven-day week. That there were seven wanderers no doubt reinforced this, each giving its name to a day of the week. In Latin the day-names were *dies Solis, dies Lunae, dies Martis, dies Mercurii, dies Iouis, dies Veneris* and *dies Saturni*. English substitutes Teutonic gods and goddesses for the Roman Mars, Mercury, Jove and Venus.

We have seen how certain solar events have been used to mark off the year and the four seasons within it. The year of four seasons seems to have begun in Mesopotamia where an alternative method of marking them was devised, perhaps by the Sumerians in the fourth millennium BC. As the Sun apparently moves eastward amongst the fixed stars, stars near the ecliptic with which the Sun is coming into near conjunction disappear after sunset (their heliacal setting) and reappear a couple of months later just before sunrise (their heliacal rising). The Sumerians (see Hartner 1965) seem to have discovered that the heliacal rising of the constellation they called *The Bull-of-Heaven* rather like our *Taurus* (or the *Bull's Jaw*, our Hyades) marked the onset of spring. Likewise the heliacal rising of *The Great Lion* (or perhaps it was *The Great Dog*), of *The Scorpion* and of *The Ibex* (largely our *Capricornus* plus our *Aquarius*) marked the onset of summer, of autumn and of winter respectively. Later in Mesopotamia eight other ecliptic constellations were added to the original four season-marking beasts, roughly one for each month.

The adoption of four seasons based on astronomical considerations was by no means universal in ancient times. The ancient Egyptians, perhaps from the early third millennium BC, recognized three seasons each of four 30-day 'months' which were no longer tied to the phases of the Moon. These three seasons were based roughly on the behaviour of the River Nile. The first season covered the usual period when the Nile was flooding, the second the usual period when the earlier flooded lands were giving forth their crops and the third when the waters dropped away and the crops were harvested. The Egyptian year consisted of 360 days (12 × 30) plus 5 days outside the 'months'.

In the first millennium BC the Aryan invaders of northern India, who perhaps passed through Persia, who developed Sanskrit as a very early Indo-European recorded language and who wrote the *Vedas* and later the *Sutras*, recognized a five-season year, each season lasting about 73 days and marked by such meteorological conditions as 'warm', 'hot', 'rains', 'frosts' and 'mists' (see O'Neil 1975).

A second practical use of a knowledge of celestial phenomena is in orientation and navigation. When travelling on land one is guided by landmarks of various sorts but the general direction of one's destination is also important. During the day one may use one's knowledge of the Sun's changing position in the sky to maintain one's bearings. On awakening the following morning one may note the Sun's rising (Latin *oriens*) point in order to get one's orientation for the coming day. When travelling at sea one may, if it is possible to hug the coasts, keep in touch with landmarks, but in the open ocean this is not possible, for example Polynesian journeys by sea in the Pacific Ocean. The height or altitude of a well-known star above the northern or the southern horizon is a good indication of the sailor's terrestrial latitude at the time. Terrestrial longitude without a very accurate timepiece is more difficult to estimate. However, noting by some rough timepiece how much earlier than the usual 40 minutes in ten days a recognized star is rising or setting would give some indication. The Polynesians, of course, used other information such as ocean current directions, bird flights and so on (see Lewis 1972 and 1974).

It may be worth noting that when the *Nautical Almanac* was introduced in 1767, the needs of navigators were stressed more than those of astronomers, though it was produced by astronomers. The *Nautical Almanac* gives the positions at specified times (at one-day or ten-day intervals) of the Sun, of the Moon, of each of the five star-like planets and of the brighter fixed stars for the coming year. The navigator, by combining the sightings of one or more of these bodies and precise times for the sightings, can work out the terrestrial longitude and latitude of his nautical or aerial vessel. From the eighteenth century onwards sufficiently precise times were given by a chronometer carried aboard ships. Prior to that only latitude could be determined with any accuracy. This is the reason so many Dutch ships left their northward turn to the

East Indies too late and were wrecked on the Western Australian coast. Today the navigator gets his time by radio signal.

A third practical motive for an interest in celestial phenomena relates to the quite early view that celestial events might have portentous or ominous significance for human affairs. Ignorant peoples have always been disturbed by eclipses of the Sun or of the Moon; something was obviously going wrong. The appearance of major comets was similarly disturbing. Once given these trains of thought people came to expect some significance for themselves in other celestial events, such as conjunctions of planets or the presence of planets in given signs of the zodiac at relevant times. The origins of such astrological interpretations of celestial events are not easily traced. The earliest evidence of astrology is probably to be found in Mesopotamia. The earliest manifestation there seems to have been judicial astrology: a celestial event is deemed to have concurrent or near future significance for the king or for the community as a whole – the king will fall ill or will vanquish a rival, or the land will suffer serious drought or will experience unusual prosperity. It seems not to have any personal significance for the individual priest, farmer or metal worker. Later, horoscopic astrology appeared. It can be found in Meso-potamia in the second or third century BC but it may have been a borrowing from Hellenistic Egypt. It predicts the fate of the individual from the disposition of the 'planets' in the zodiac at the time of his birth. Such horoscopic astrology spread at the beginning of the Christian era widely in the Roman Empire, especially around its periphery in the newly conquered provinces. It was believed by the Greeks and Romans who commented on it at the time to have come from Egypt. It spread perhaps from late Hellenistic Mesopotamia or from the early Roman Near-East into India, where it has flourished up to the present day, and into a number of neighbouring cultures. It may be worth remarking that while few newspaper astrologers in the West would be capable of reading the U.K.–U.S.A. *Nautical Almanac* produced mainly for sea and air navigation, the Indian *Nautical Almanac* provided additional data of interest primarily to astrologers who were engaged in casting horoscopes. For the development of astrology see Pingree 1974a and Sachs 1952a.

Of all of this early concern with celestial events one may ask what is scientific astronomy or at least proto-scientific astronomy and what is only some precursor interest. In the Genesis story, God paraded each of his creatures in front of Adam and invited him to give each a name. Adam probably had some taxonomic problems which could only be resolved later by much tedious observation and thought.

Distinguishing and naming may be an important precursor activity to the development of science but it is not in itself science. The identification and naming of stars and constellations and the distinction between planets and fixed stars is at best a pre-scientific activity even though it may be a necessary prior activity. If one is to find regularities in the various celestial events one must be able to identify them and preferably to name them. The early Greek attribution of different names to Mercury and to Venus when each were morning and evening stars is an example of a taxonomic problem.

Pre-scientific astronomical activity, involving observation, classification and the noting of broad regularities, ultimately gave rise to primitive or proto-scientific astronomy, which later developed into scientific astronomy. The dividing lines are not easy to specify. In pre-scientific astronomy the naming and classifying were entirely qualitative and almost completely idiosyncratic. The next step was to look for regularities of recurrence in at least rough quantitative terms. Later, mathematical systems were produced in order to enable prediction and perhaps explanation; here scientific astronomy began.

Western astronomy had its beginnings probably in Mesopotamia perhaps in proto-scientific form in the first half of the second millennium BC. By the end of the first half of the first millennium BC, it was approaching a genuinely scientific state in which arithmetico-algebraic analyses were used. Much of this later Babylonian mathematically based astronomy passed over to Greece where geometrical systems were developed (see Neugebauer 1969 and 1975). This knowledge passed into India first from the Hellenized Middle East and later from the Roman Empire in the East. Astronomy in the West was virtually dead when the Arabs imported Indian astronomy. The Arabs also found and translated Greek texts or Syrian translations of Greek texts. They engaged in observational work over several centuries. Towards the end of the Middle Ages the Arabian astronomers were very active in Spain, whence their texts percolated into Christian

Europe, firing the revival of Western astronomy.

The Sumerian, Akkadian, Greek, Latin, Sanskrit and Arabic names for the seven planets seem to be unrelated (see Chapter 12), except where there are given names of gods and goddesses playing similar roles in the hierarchy of deities. On the other hand the names of the twelve zodiacal constellations and the related but somewhat later signs of the zodiac show a marked continuity from the Sumerian names to the Arabic (see Chapter 12).

The geometrical astronomical theory which arose in ancient Greece persisted in India and then in Arabia, had a revival in Europe and experienced its finest flowering at the hands of Copernicus who through opening the door for modern astronomy may best be regarded as the last great astronomer in the ancient tradition. The following chapters will trace this tradition from the Babylonians. There are two appended chapters, one on the probable prehistoric astronomy which may have been responsible for many megalithic monuments in Britain and Brittany, and one on Chinese astronomy which was largely outside the tradition examined in the main chapters.

CHAPTER TWO

EARLY BABYLONIAN ASTRONOMY

Groups of people who spoke Akkadian, an east Semitic language, settled first in Babylonia on the Euphrates and later in Assyria on the Tigris from about the end of the third millennium BC onwards. Their southern neighbours in Mesopotamia were the Sumerians who themselves were migrants, perhaps from the northeast in Persia. Some of the Sumerian early myths and legends refer to mountains and forests, both non-existent in lower Mesopotamia but characteristic of western Persia. The Sumerians spoke an agglutinative language unrelated to any other known language and they developed a method of writing it in wedge-shaped (cuneiform) symbols on wet clay tablets which were subsequently hardened by sun-drying or by baking. The Sumerians who settled in lower Mesopotamia perhaps in the fifth or early fourth millennium BC may have been originally hunters but they became herders of sheep, goats and large cattle and cultivators of grain such as wheat and barley and of fruits such as dates and grapes, irrigated by water levered in buckets from the river. They became skilled artisans in spinning and weaving, pottery, the working of leather, and the shaping of wood, stone and metal, all of the latter three being in the main imported. They made bronze from imported copper and tin. At their peak they organized themselves in small walled city-states under a king who regarded himself as the vicar of the local god of his city. There was considerable division of labour – the herders, the farmers, the artisans of various sorts, the warriors headed by the king, and the priests who were the scribes and the scholars.

The Sumerians came to engage in extensive trade in order to exchange goods which they had in surplus to their own needs, such as grain, textiles, hides and various other maufactured goods in wood, metal and stone, for goods which were in short local supply, mainly raw materials such as timber, stone, especially gemstone such as lapis lazuli, and metals, especially copper and tin for the making of bronze. Their merchants spent much time in foreign cities far afield in order to cultivate this trade. Their invention of writing was probably prompted by the need to have a record of trade transactions as well as of local administrative activities involving taxation and inheritance. Written records were extended to prayers and rituals and to their rich mythology including what is probably the model for the Biblical flood story. They too had stories of creation but they were quite different from those in Genesis (see Kramer 1959, Hooke 1963, Oppenheim 1964 and Gordon 1967).

The Akkadian-speaking Babylonians and Assyrians borrowed all of these things from the Sumerians whom the Babylonians subjugated and later assimilated. The Babylonians and Assyrians tended to combine the semi-independent city-states of the Sumerians into empires, great or small. For hundreds of years the Babylonian and Assyrian empires, often spreading well beyond Mesopotamia, the land between the rivers, were in conflict, though one sometimes subjugated the other in enforced suzerainty and sometimes the two united in roughly joint equality. Akkadian speakers adapted the Sumerian cuneiform symbols which recorded whole words in order to record the several syllables of their words. They also preserved the Sumerian language for ritual and technical purposes much as medieval Europe preserved Latin (with modifications) for similar purposes. As the scholars have not been able to trace the use of Sumerian as successfully as the European medievalists can trace the use of an evolving medieval Latin, it is difficult to say whether some notions expressed n the preserved Sumerian language are in origin Sumerian or are Babylonian notions recorded in anachronistic

Sumerian. There are few, perhaps no, original Sumerian tablets with a significant astronomical reference, though there are many using Sumerian words but apparently written in Babylonian or Assyrian imperial times.

It is likely, however, that many Mesopotamian pre-scientific astronomical notions and distinctions were Sumerian. The distinction between the so-called fixed stars and the seven wanderers (*bibbu* or wild sheep to the Babylonians, later called *planetoi* or wanderers by the Greeks) may have been made by the Sumerians. The Sumerians seem to have named many of the bright fixed stars and many constellations of stars. Four constellations, as we have seen, were recognized by 3000 BC (Hartner 1965). As they reappeared before sunrise, their heliacal rising, after a period of invisibility, they marked the onset of spring, summer, autumn and winter. The heliacal rising of the *Bull-of-Heaven* or the *Bull's Jaw*, whose head was formed by the vee-shaped open cluster, the Hyades, with the bright star Aldebaran as a shining eye, marked the onset of spring *circa* 2500 BC. Likewise the heliacal rising of the *Great Lion* marked the onset of summer, that of the *Scorpion* the onset of autumn and that of the *Ibex* the onset of winter.

To these four season-marking constellations, eight others were added, presumably making one for each of the twelve months in an ordinary year. The *Great Twins of Heaven* and the *Worker in the River-bed* (or the *Crab*) were inserted between the *Bull-of-Heaven* and the *Great Lion*; the *Furrow* and the *Weighing Scales* were inserted between the *Great Lion* and the *Scorpion*; the *Ibex* was divided into the *Soldier* (or *Arrow-shooter*) and the *Goat-fish*, and the *Great Man-of-Heaven*, our *Aquarius*, depicted as emptying an urn of water, was added after them; next were added the *Fish tails* (a pair of fish tied with a cord at the tails) and the *Hired Farm Labourer*, our *Aries*.

The Mesopotamian *Bull-of-Heaven* shared with our *Taurus* the cluster Hyades as its head but in other repects it was different (see Fig. 2.1).

Late in Babylonian times the *Furrow* became a winged fertility goddess standing in a furrow, holding a stalk and ear of wheat in one hand and a date palm frond in the other – this was the Byzantine and medieval European representation of *Virgo*. *Lu.hun.ga* was in Sumerian the *Hired Farm Labourer*, in Akkadian this became *agru*. The Sumerian name was often abbreviated to *Lu*

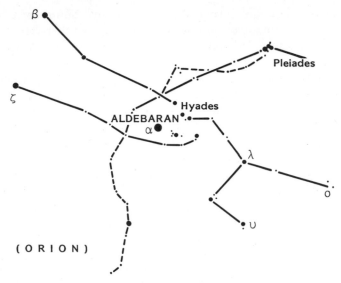

Fig. 2.1 *The main stars, joined by continuous lines, making up our constellation* Taurus. *The Babylonian* Bull-of-Heaven *had only the vee-shaped cluster Hyades in common, the stars joined by broken lines constituting the crumpled horns.*

or to *Hun*. The equivalent of *Lu* as a separate word in Akkadian was *immeru*, a sheep. Hence it seems that all the Graeco-Roman zodiacal names began in Babylonian times.

Because so much technical, ritualistic and legendary material was written in Sumerian by Babylonian scribes, it is difficult to say what was Sumerian and what Babylonian in terms of discovery, recognition or invention. There is an Epic of Creation (see Langdon 1923), almost certainly of Sumerian origin but with clear Babylonian modifications. It says amongst other things that Marduk, the principal god of Babylon, established the months and allotted three stars to each. Marduk is a Babylonian and not a Sumerian god, and so if the epic is Sumerian there has been a change of *dramatis personae* in the myth as preserved for us. A fairly early Babylonian tablet, or more correctly a copy of a fairly early tablet, from say the late second millennium BC, names the three stars for each month and locates each in one of three ways on celestial bands (see van der Waerden 1949).

Anu's way seems to have been centred on the celestial equator, Enlil's way to the north and Ea's way to the south. Anu, Enlil and Ea were the gods of the air, of the heavens, and of the waters and the earth respectively. The Sun was said to spend three hot, dry months in Enlil's way, two sets of three stormy and windy months in Anu's way and three cold watery months in Ea's way. It is worth noting that the three winter zodiacal constellations in the second millennium BC were the goat-fish, the water-pourer and the two fishes tied by their tails, all having a reference to water. It is also worth noting that the three ways or celestial bands are oriented to the celestial equator whereas later Babylonian astronomy used an ecliptic frame of reference. A further point needs to be made about this tablet. In three cases one of the stars belonging to a month bears the name of a planet and in a rather greater number of cases the named fixed star or set of stars was not in fact in the celestial band specified. These two peculiarities may indicate that the copyists who wrote the tablet which we have did not correctly read the tablet being copied (because it was abraded or otherwise defective) or that the compilers of the original were poor astronomers (see Fig. 2.2)

Gössmann (1950) lists 407 celestial objects for which we have recorded Sumerian or Akkadian names (and sometimes both). These include the Sumerian *Dur-an* and Akkadian *Rikis.same* for the Milky Way; the Sumerian *Ud* and Akkadian *Samas* for the Sun; the Akkadian *Sin* for the Moon; several names for Mercury, *Ubu-idim-gud-ud* or *Gud-ud*; for Venus, *Dil.bad* or *Nin.si-an.na*; for Mars, *Salbatanu* or *Sanumma*; for Jupiter, *Mul-babbar*, *Sag.nae-gar* or *Marduk*; and for Saturn, *Sag.us*, *Genna* or *Udu.idim*. There are 58 names for constellations, many similar in pattern to ours but some markedly different – thus *Ur-gu-la* (great dog or lion) is very like our *Leo*, whereas our *Canis Major* was seen as a bow, *Ban*, and a separate arrow, *Kak-si-di*, the latter being tipped by our star Sirius. This leaves 341 names for individual stars such as our Aldebaran (*Ne-gi-ne-gar*), our Antares (*Lisi-gun*), our Canopus (*Nun-ki*), a name now given to a different star, and so on, or for asterisms such as the Hyades (*Gis-da*), the Pleiades (*Mul-mul* or *Zappu*) and Praesepe (*Nangar*). This is not a large proportion of stars visible to the naked eye from Babylon but it is a substantial number for their Mesopotamian names to have been preserved on the clay tablets.

We know that a great deal of tablet copying went on. For

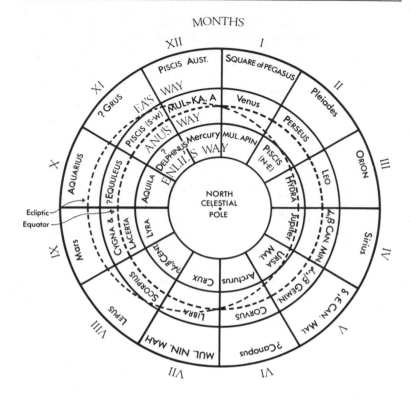

Fig. 2.2   The early Babylonian 36 stars. They date from the latter part of the second millennium BC. They were allocated to three bands or ways (Ea's, Anu's and Enlil's) oriented to the celestial equator. Four of the star-like planets were included amongst them although they could not serve as a month's marker star. In month X, Aquarius, Equuleus and Aquila were in the right order, south to north, but in the epoch concerned the first and third were only on the edges of Anu's way. In month II the three 'stars' were markedly out of order: the Pleiades were in the middle of Anu's way, Perseus stretched from Anu's into Enlil's way and the northeastern member of Pisces was in the southern part of Anu's way.

example Assurbanipal, a king of Assyria in the seventh century BC, had records from all over Mesopotamia copied for the great library he established. Though a sun-baked or fire-baked clay tablet is very durable if buried in the soil, it is prone to break and to have its surface chip if frequently handled. Thus after a time a copy became necessary. If the copyist did not understand the partially illegible original or if he were careless and inattentive, he could easily introduce errors.

This outcome is well illustrated by a set of fragments of tablets copied mainly in the eighth and seventh centuries BC, copies made in many cases for Assurbanipal's library in Nineveh. When they were reported in 1928 by Langdon and Fotheringham there were two major fragments, a pair consisting of two broken but rejoinable parts of a tablet, a third fragment was less extensive and four others were quite minor in the data they provided. The information presented consisted of (i) the dates of last visibilities of Venus before superior or inferior conjunctions for most of the 21 regnal years of some unnamed king, (ii) the dates of first visibility after superior or inferior conjunction, (iii) the interval of invisibility between last and next first visibilities, and (iv) sometimes the interval of visibility between first visibility and next last visibility. Typical entries are:

a   If on the 21st day of Abu [the fifth month] *Nin.si-an.na* [Venus] disappeared in the east, remaining absent from the sky for two months and eleven days, and on the second day of Arasamu [the eighth month] she is seen in the west there will be rains in the land and desolation will be wrought.

b   If on the 26th day of Ululu [the sixth month] *Nin.si-an.na* disappeared in the west remaining absent from the sky for eleven days, and was seen on the 9th day of the Second Ululu [the extra month let into a full or thirteen-month calendar year] in the east, the heart of the land will be happy. (Langdon and Fotheringham 1928)

The first of these two entries centres on an invisibility around the superior conjunction of Venus, an invisibility ordinarily from one month 25 days to two months ten days, and the second on an invisibility around the inferior conjunction, ordinarily from one day to eighteen days (see Fig. 2.3). These periods of invisibility

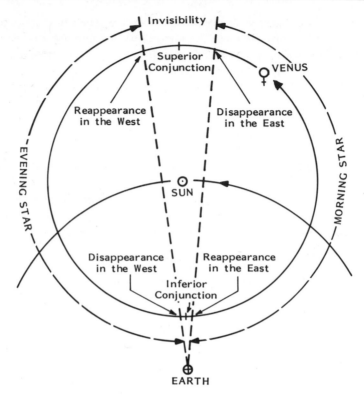

Fig. 2.3 *The periods of visibility (when clear of the light of the Sun) and of invisibility of Venus.*

could be lengthened by a few days if poor viewing conditions, for example clouds, mist or dust near the horizon, prematurely obscured or unduly delayed the sighting of Venus in the western or eastern sky, and the period of visibility could be similarly shortened.

The two entries quoted are typical of most but not all of the entries, namely a date of last visibility, an interval of invisibility, a date of first visibility and an omen, which was sometimes followed by the subsequent interval of visibility. Though the record appears to be for the 21 regnal years of some king, the only clue as to who the king might have been is given by the replacement of the omen in the eighth year by the statement 'the year of the golden throne'. Regnal years were sometimes referred to by such means as 'the *x*th year of King *Y'* but sometimes they were referred to by means of some remarkable event. We know from independent evidence that the eighth regnal year of Ammisaduga, who reigned for 21 years, was called 'the year of the golden throne'. Hence presumably these Venus records refer to the reign of Ammisaduga, the fourth Babylonian king after Hammurapi. Reiner (1975) has found, mainly in the British Museum, many fragments additional to those known to Langdon and Fotheringham. Two, one a pair of joinable fragments, gave extensive information, whereas nine others give only limited information; some of the latter may be fragments of other tablets known to Langdon and Fotheringham or only to Reiner.

In 21 Babylonian calendar years, most of which would have had twelve months of 29 or 30 days, but probably seven or eight of which had thirteen months, there would have probably been 26 last visibilities and 26 first visibilities (each divided equally between such events prior to or after inferior and superior conjunction), 26 periods of invisibility and 25 periods of visibility. We have other evidence (see Reiner, p. 23) that some seven of the 21 years of Ammisaduga were thirteen-month years.

On the many fragments analysed by Reiner, there are 24 references to dates of last visibilities, 24 to periods of invisibility, 23 to dates of first visibilities and 22 to periods of visibility. Many of the references are repeated in the several fragments, thus eleven last visibilities are duplicated, two triplicated, one quadruplicated and one is even quintuplicated. Usually the several dates in such a multiple record for a given event coincide or differ by only a few days. Thus of the 49 last and first visibilities, fifteen have a single entry, thirteen with two to four entries give identical dates, nine with two to three entries give dates differing by one to four days, eight give dates differing by eight to seventeen days and one event with two entries has a discrepancy of 21 days. Two discrepancies are of two months and one of three months. This last, a triple, could have occurred because the scribe who made the copy misread the month name, which may have been abraded on the tablet being copied. This hypothesis is supported by an alleged period of invisibility around inferior conjunction of nine months four days, which is a complete absurdity. The date of last visibility is given as the eleventh day of the third month and the date of the first visibility in that year, Year 9, is given as the fifteenth day of the twelfth month. The intervals between last visibilities before inferior conjunctions average nine months fourteen days. If we come forward from the relevant last visibility in Year 8 and backward from the relevant last visibility in Year 11 by that interval the suggestion is that the scribe wrote the name of the third month instead of the name of the twelfth month. In that event the period of invisibility would be four days, which fits well within the limits of the period of invisibility around inferior conjunction.

Allowing a few extra days for a period of invisibility, 20 of the 24 claimed are within the astronomical limits (and three of those outside can be plausibly attributed to miscopying); sixteen of the 22 periods of visibility are within the astronomical limits of eight months and a few days – four are under eight months and two over nine months (most plausibly attributable to miscopying of the month name).

Eliminating a few recorded dates with obvious gross errors, a number of values for Venus periods may be calculated from the tablets of Ammisaduga. For example, the recorded dates imply a synodic period for Venus of about 583 days (the best modern value is a mean of 583.9 days with a maximum of about 590 and a minimum of about 578) and periods of invisibility averaging 9.4 days and 65.7 days, which are close to the best modern values. All in all, this is a remarkable piece of early astronomical observation carried out probably in the seventeenth or sixteenth century BC. It would be surprising if it were unique in its own time; more likely it is unique in its survival and recovery.

As we shall see in respect of Neo-Babylonian records it is possible to establish in modern terms the dates and even times within the day of reported phenomena and to compare the reported dates and times. Unfortunately the scholars have not been able to establish a convincing chronology for the First Babylonian Dynasty. The dates in the Venus tablets have been used in attempts to establish the chronology of the reign of Ammisaduga. Unfortunately the cycle which is approximately indicated in the Venus tablets repeats every 40 or 50 years. Various scholars have argued for 1920 BC, 1701 BC, 1645 BC and 1581 BC as the first year of Ammisaduga (see Weir 1972). If we had independent evidence of the dates, in modern terms, of Ammisaduga's reign we could better assess the accuracy of the dates and intervals in the Venus tablets.

Evidence of further astronomical work after the Venus tablets and the 36 stars in Enlil's, Anu's and Ea's ways is missing until about the middle of the eighth century BC. Ptolemy in his *Syntaxis mathematike* (better known by the title of its Arabic translation, *Almagest*), written in the second century AD, claimed that he had Babylonian eclipse records from the time of King Nabonassar, who was enthroned in 747 BC. We have some records from the seventh century BC which indicate that the Babylonians were making methodical observations and were finding patterns which enabled them to make predictions of dates of eclipses or possible eclipses and of dates of planetary phenomena such as disappearances and reappearances, before and after conjunction with the Sun, of the beginning and end of the retrograde phases (apparent westward motion) and, in the case of Mars, Jupiter and Saturn, of opposition with the Sun in the middle of the retrograde phase. Later, in addition to dates, the position in the sky is given in terms of so many degrees of longitude within a sign of the zodiac (a concept introduced only about 500 BC) or in terms of some 31 reference stars near to the ecliptic. As the latter reference system is cruder than the signs of the zodiac it may be earlier but that is only a conjecture. The reference stars, often called normal stars, were used contemporaneously with the signs of the zodiac for some centuries in the second half of the first millennium BC.

The reference stars can be confidently identified with stars known at present and their longitudes and latitudes relative to the ecliptic can be calculated for relevant epochs. In 301 BC their latitudes ranged from $7°.0S$ to $9°.7N$ of the ecliptic with a mean $1°.8N$ (see Table 2.1). Of the 31 differences in longitude, ten are 5° or less, eight are between 5° and 10° and seven are between $10°$ and 15°; these 27 differences have a mean of $5°48'$. The remaining six differences are $16°36'$, $21°12'$, $22°18'$, $31°41'$, $42°42'$ and $63°24'$. This unevenness in the longitudinal spacing of the reference stars has two serious defects. First, the Moon advances on average by some $13°20'$ per day (sometimes overtaking several reference stars in one day and in the worst case only one in about four days): Mars moves forward on average almost 2° in a day, ignoring its retrograde phase and so on. It is as though the celestial milestones were on a few occasions one mile apart but on others up to 63 miles apart with their numbering taking no account of these inequalities of the units. Second, such unequal units make arithmetical summing impossible.

Table 2.1 The 31 Babylonian reference stars with their longitudes and latitudes for 301 BC

Star	Long. (degrees)	Lat. (degrees)	Star	Long. (degrees)	Lat. (degrees)
Beta Ari.	2.0	+8.4	Rho Leo.	124.4	0.0
Alpha Ari.	5.7	+9.9	Theta Leo.	131.4	+9.7
Eta Tau.	28.0	+3.8	Beta Vir.	144.7	+0.7
Alpha Tau.	37.8	−5.6	Gamma Vir.	158.5	+3.0
Beta Tau.	50.6	+5.2	Alpha Vir.	171.9	−1.9
Zeta Tau.	52.8	−2.5	Alpha Lib.	193.2	+0.6
Eta Gem.	61.5	−1.2	Beta Lib.	197.4	+8.8
Mu Gem.	63.3	−1.1	Delta Sco.	210.6	−1.7
Gamma Gem.	67.1	−7.0	Beta Sco.	211.2	+1.3
Alpha Gem.	78.4	+9.9	Alpha Sco.	217.8	−4.3
Beta Gem.	81.6	+6.5	Theta Oph.	229.4	−1.5
Theta Cnc.	93.8	−1.0	Beta Cap.	272.1	+4.9
Gamma Cnc.	95.6	+3.0	Gamma Cap.	289.7	−2.3
Delta Cnc.	96.7	0.0	Delta Cap.	291.5	−2.2
Epsilon Leo.	108.7	+9.5	Eta Psc.	354.9	+5.2
Alpha Leo.	118.0	+0.4			

It is difficult to understand this uneven spacing of the reference stars. In general where the spacing is least there are more bright stars than when it is greatest. However, in the former regions relatively dim stars were used when brighter ones were passed over, whereas in the latter region there was a short supply of very bright stars but an ample supply of moderately dim stars, say third or fourth magnitude, a concept which will be unfolded later.

There are two kinds of later Babylonian astronomical records which had their origin prior to 500 BC, the earliest date for the establishment of genuinely scientific astronomy.

First are what have been called diaries, usually covering half a year, that is six or seven months, for which we have an example datable to 651 BC, though they probably went back to Nabonassar's reign, that is to about 747 BC. Second, there are goal texts, from about 250 BC, although they seem to have Babylonian and Assyrian predecessors of a much earlier date, or at least to use a principle discovered in earlier times (see Sachs 1948).

Though the diaries seem to have dated from the first half of the first millennium, the most numerous and complete date from the second half of the first millennium, hence an account of them will be postponed to the next chapter.

The goal texts are based on the fact that after certain so-called resonance periods planetary phenomena, such as disappearances, reappearances, first and second stations (beginning and end of the retrograde phase of apparent motion), and in the case of Mars, Jupiter and Saturn, the oppositions, repeat in a given pattern of intervals. Thus a fairly early tablet, probably fifth or even sixth century BC, though it may be stating an earlier discovery, states in effect 'for the dates of the phenomena of Venus go back to the record of eight years before and subtract four days', 'for Mars go back 47 years and add twelve days'. The suggested addition of twelve days in the case of Mars is some ten days too many; perhaps the value was established before the Babylonians had a regular scheme for adding the thirteenth month to seven years in a nineteen-year cycle, that is before about 500 BC, or perhaps the numeral 2 was miscopied as 12.

In the later goal texts the interval from the earlier year was set out planet by planet and the corrections specified. Jupiter and Mars had two paradigmatic years and Mercury had three, thus the complete set was:

Saturn	59 years	(subtract 6 days)
Jupiter	71 years	(same date)
Jupiter	83 years	(subtract 13 days or add 17 days)
Mars	47 years	(add 2 days)
Mars	79 years	(add 7 days)
Venus	8 years	(subtract 4 days)
Mercury	6 years	(add 14 days or subtract 16 days)
Mercury	13 years	(subtract 4 days)
Mercury	46 years	(subtract 1 day)

Where more than one rule is used for a given planet, it is because one reproduces one value of a phenomenon better than another, say longitude rather than date. These goal texts and their more cryptic predecessors stating the rule without applying it were almost certainly based on the diaries.

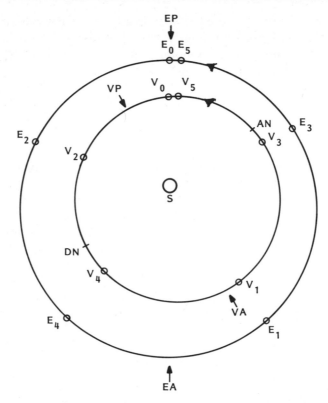

Fig. 2.4   *Earth and Venus have slightly excentric orbits and so speed up around their perihelia (EP and VP) and slow down around their aphelia (EA and VA). When Earth is at* E_0, *Venus is at* V_0 *and in conjunction with the Sun: some 584 days later Earth will be at* E_1 *having made almost 1.6 orbits and Venus will be at* V_1 *having made almost 2.6 orbits and be in conjunction with the Sun again. After further periods of about 584 days Venus will be in conjunction with the Sun at* V_2, V_3 *and* V_4. *At the fifth successive conjunction (* V_5 *) after almost eight years, Earth and Venus will be back almost where they were at the initial conjunction.*

These resonance periods result from the excentricity of the Earth's orbit and that of the orbit of the planet concerned, though the Babylonians almost certainly did not know this. The resonance period of Venus is represented diagrammatically in Fig. 2.4. The orbit of Venus departs very slightly from a circle and that of the Earth slightly more. At E_0 and V_0, Venus is in conjunction with the Sun and both Earth and Venus are at or approaching perihelion and hence travelling fast. At the next conjunction, E_1 and V_1, Earth will have made 1.598 revolutions and Venus 2.598 revolutions and both will be near aphelion and travelling slowly. After 7.99 revolutions by Earth and 12.99 by Venus, they will be back at E_5 and V_5 in conjunction, as seen from the Sun in essentially the same circumstances as they were almost eight years before.

One other regularity which was probably discovered before 500 BC, although our main record is early third century BC (see Table 2.2), has to do with eclipses, almost certainly lunar eclipses. Compared with ephemerides (tables giving positions) of the second century BC, it has an archaic flavour and may well derive from an earlier period than the one it covers. The remains of one tablet broken off on both sides contains columns of dates in sets of 3, 8, 7, 8, 8 and 4, spaced at six months within each set but with five months between each set. The remnant runs from the second half of the 31st regnal year of Artaxerxes II, i.e. 373 BC, to the 35th year in the Seleucid era, i.e. 277 BC. The first four kings in this list were Persian, whereas the next five were Greek. This tablet must have been written after 280 BC because it could not have been known until about that date that regnal years of individual kings would be replaced by years in the Seleucid era, which began in 311 BC.

Each column in the tablet has 38 month names and eighteen regnal year numbers. No days in the month are given, presumably because lunar eclipses occur in the middle of Babylonian calendar months which are basically synodic. If we add the three dates at the head of each column to the four at the bottom of the preceding column, we have as set out in Table 2.2, sets of 8, 7, 8, 8 and 7 dates which are six synodic months apart within themselves and which are separated by five synodic months. This yields an eclipse cycle of 223 synodic months. Each eclipse in fact tends to be repeated after 223 synodic months. We can understand the basis of this regularity but it is most likely that the Babylonians discovered it as a brute fact for which they had no explanation.

A lunar eclipse occurs when the Moon is full and is near a node, one of the two points where the Moon's orbit crosses the ecliptic (see Fig. 1.3). On such occasions the Moon passes through the

Table 2.2   Part of a Babylonian fragment recording possible lunar eclipses

Art.	33		II				XII		40		IV
			VIII		37		VI				X
	34		II				XII		41	*int.*	III
			VIII		38	*int.*	V				IX
	35	*int.*	I				XI		42		III
			VII		39		V				IX
	36		I				XI		43		III
			VII								IX
	44	*int.*	I				XI		5		III
			VII		2		V				IX
	45		I				XI		6	*int.*	II
			VII		3	*int.*	IV				VIII
			XIIa				X		7		II
	46		VI		4		IV				VIII
			XII				X		8		II
Och.	1		VI							*int.*	VII
			XII		12		IV		16		II
	9		VI				X				VIII
			XII		13		IV		17	*int.*	I
	10		VI				X				VII
			XII		14	*int.*	III		18		I
	11	*int.*	V				IX				VII
			XI		15		III				XIIa
							IX		19		VI
			XI		2		III		4		I
	20		V				IX				VIa
			XI	Dar.	1		III				XII
	21		V				IX		5		VI
			XI		2	*int.*	II				XII
Ars.	1	*int.*	IV				VIII	Alex.	1		VI
			X		3		II				XII
							VIII				

2 int.	IV		6	II	int.	XII
	X			VIII	3	V
3	IV		7	II		XI
	X			VIII	4	V
4	IV	Phi.	1 int.	I		XI
	X			VII	5	V
5 int.	III		2	I		XI
	IX			VII		

In each of the three columns the first sub-column gives the regnal year – Art. is Artaxerxes II, Och. is Ochus, Ars. is Arses, Dar. is Darius, Alex. is Alexander the Great and Phi. is Philip. The month number is given in the third sub-column – VIa and XIIa indicate a second (i.e. intercalated) sixth and twelfth month respectively. *Int.* indicates that there has been an intercalated month since the month given in the preceding line.

Earth's shadow (see Fig. 2.5). The Moon is full on average every 29.53059 days (the synodic month) and makes a double crossing (an ascending node and a descending node) of the ecliptic on average every 27.21222 days (the draconic month). Thus about every six synodic months (6 × 29.53059 days or 177.18354 days) and every six and a half draconic months (6.5 × 27.21222 days or 176.87943 days) the Moon will be full and near a node. The Moon moves eastwards on average at $13°.176$ per day so that after six synodic months and approximately six and a half draconic months the full Moon will have moved about $2°.689$ closer to a node or by that amount past it on average.

Thus for some three to five successive six-month intervals the full Moon will be within 5° to 6° of a node and there will be a total eclipse. On one to three surrounding six-month intervals the full Moon with be within 10° to 12° of a node and there will be a partial umbral eclipse. Beyond that separation there will be one or two penumbral eclipses and then a jump of five months to the next node which the full Moon has been approaching. This sequence does not go off with complete regularity because the Moon's eastward progression in respect both of returning to a given node and of catching up with the Sun has a variable velocity. Thus the actual angular distances of the full Moon from a node may be almost 2° greater or less than those suggested on the basis of mean values. For comparison, a series of lunar eclipses in this century is set out in Table 2.3

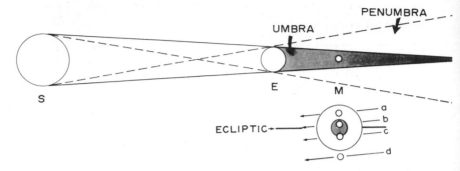

Fig. 2.5 *The shadow of the Earth. It is a double shadow centred on the plane of the Earth's orbit. The umbra is dense and the penumbra diffuse. If the Moon when at opposition is within about 18° of a node, it may enter the penumbra, resulting in a penumbral eclipse (a). If it is within about 12° of a node it will partially enter the umbra, resulting in a partial eclipse (c). If it is within about 6° of a node it will completely enter the umbra, resulting in a total eclipse (b).*

Were only the synodic and the draconic months involved in the occurrence of lunar eclipses the patterns within each set of seven or eight eclipses six synodic months apart would be more regular than they are. The third factor involved is the anomalistic month which averages 27.55455 days between successive perigees, when the Moon is nearest the Earth. When the Moon is near perigee it will be moving faster than when it is near apogee. The draconic and anomalistic months differ on average by some 0.34 days which over 6.5 draconic months produces a difference of some 2.2 days or about 27° difference in the Moon's longitude, thus producing a variation in the pattern of penumbral, partial and total lunar eclipses, as can be seen in Table 2.3.

The following near equalities are important: 223 synodic months of 29.53059 days equal 6585.32 days; 242 draconic months of 27.21222 days equals 6585.35 days and 239 anomalistic months of 27.55455 days equals 6585.54 days. Thus after 223 synodic months, the so-called Saros cycle, the alignments of the Moon and the Earth with the Sun are almost back where they were at the beginning of this period, which constitutes each column in Table 2.3. The Babylonians discovered this cycle perhaps not later than the first half of the first millennium BC; it was the English

Table 2.3   Dates and types of lunar eclipses from 16 October 1940 to 24 March 1959

Date of eclipse	Type	Months since last	Date of eclipse	Type	Months since last
6 Oct. '40	Penumbral	6	26 Sept. '50	Total	6
3 Mar. '41	Partial	5	23 Mar. '51	Penumbral	6
5 Sept. '41	Partial	6	17 Aug. '51	Penumbral	5 ⎫
3 Mar. '42	Total	6	15 Sept. '51	Penumbral	1 ⎭
6 Aug. '42	Total	6	11 Feb. '52	Partial	5
9 Feb. '43	Partial	6	5 Aug. '52	Partial	6
5 Aug. '43	Penumbral	6	29 Jan. '53	Total	6
9 Feb. '44	Penumbral	6	26 July '53	Total	6
6 July '44	Penumbral	5 ⎫	19 Jan. '54	Total	6
4 Aug. '44	Penumbral	1 ⎭	16 July '54	Partial	6
9 Dec. '44	Penumbral	5	8 Jan. '55	Penumbral	6
5 June '45	Partial	6	5 June '55	Penumbral	5
9 Dec. '45	Total	6	29 Nov. '55	Partial	6
4 June '46	Total	6	24 May '56	Partial	6
8 Dec. '46	Total	6	18 Nov. '56	Total	6
3 June '47	Partial	6	13 May '57	Total	6
8 Nov. '47	Penumbral	6	7 Nov. '57	Total	6
3 Apr. '48	Partial	5	4 Apr. '58	Penumbral	5 ⎫
8 Oct. '48	Penumbral	6	3 May '58	Penumbral	1 ⎭
3 Apr. '49	Total	6	27 Oct. '58	Penumbral	6
7 Oct. '49	Total	6	24 Mar. '59	Partial	5
2 Apr. '50	Total	6			

The first, third and fifth series have eight eclipses which are six months apart, plus an additional eclipse only one month before the regular eclipse, and the second and fourth series have seven eclipses which are six months apart. Only five months elapse between the last eclipse in one series and the first in the next.

astronomer Halley who mistakenly thought that they applied the word Saros to it.

It is a matter of some puzzlement as to how the Babylonian astronomers in their earlier phase discovered the eclipse cycle reported. A few thoughts, however tentative, may provide some illumination. Perhaps in a sequence of a few sets of six synodic months a few lunar eclipses would have been observed and inferences drawn about some intermediaries which had occurred below the horizon – the rising full Moon opposite the Sun after sunset on one day but ahead of it on the next could suggest an unobserved intervening eclipse. Filling in the gaps for unobserved eclipses because they had occurred below the horizon or were penumbral would yield an interpolated and extrapolated series of possible eclipses six synodic months apart. Similar series could be established for an earlier or a later period but where these six-month series interlocked there was an interval of only five months. I suggest that though the eclipse cycle presented in Table 2.3 can be generated on 'rational' grounds which have been stated, it was probably generated in the first millennium BC in Mesopotamia on completely 'empirical' grounds. Sequences of observed eclipses and intermediate eclipses which were assumed to have occurred below the horizon were six months apart. The next sequence which also consisted of eclipses observed or inferred to be six months apart was seen to involve the dropping of a month where one sequence intersected with the next sequence. It was only later that an explanation of these patterns could be provided. By the time the tablet which I have been discussing was written the Babylonian astronomers were on the eve of providing an analytic explanation.

It is important to emphasize that many Babylonian astronomical records which may appear to be reports of observations are in fact reports of predictions. Even in the diaries there are many entries accompanied by such comments as 'it passed by' or 'it could not be observed' or 'it was expected on such and such a date but it occurred on another specified neighbouring date', indicating that the entry is a prediction.

LATE BABYLONIAN ASTRONOMY

*E*arly Babylonian astronomy, as
we know it in the second millennium and the first half of the first
millennium BC, was pre-scientific or at best proto-scientific. By
about 500 BC it was approaching a genuinely scientific status. Two
factors contributed. The first was the accumulation of observational records as in the diaries, perhaps from about 750 BC although
the earliest diary we have is from about 650 BC. In their developed
form (say from the fourth century BC) the diaries stated month by
month for a half year the following information (see Sachs 1948):

i   The length of the previous month, 29 or 30 days, and the time
interval between sunset and the setting of the first visible crescent
Moon which marked the beginning of the new month. Time
intervals were given as time degrees, 1° equalling in our terms 4
minutes, a relationship preserved in our hours and minutes of
Right Ascension.

ii   Around full Moon, the intervals between moonset and
sunrise, between sunrise and moonset, between moonrise and
sunset and between sunset and moonrise (when those events
occurred in the order stated).

35

iii   The date of the last visible crescent Moon and the interval from moonrise to sunrise on that occasion.

iv   Lunar and solar eclipses, both those observed and those predicted but not observed and said to have 'passed by'. The time when a lunar eclipse began, the duration from beginning to greatest magnitude and the degree of maximum magnitude, that is, whether total or, if partial, what proportion of the Moon was in the umbra.

v   For Mars, Jupiter and Saturn, the dates of reappearance, first station, first visible rising after sunset (that is, after opposition), second station and disappearance, the sign of the zodiac occupied at first and last visibility. For Mercury and Venus, dates and sign of the zodiac for reappearance and disappearance as morning and as evening stars.

vi   Dates of equinoxes and solstices (obviously calculated rather than observed for they are evenly spaced); dates of the heliacal risings of Sirius apparently used for the calculation of the equinoxes and solstices (if so, this ignored precession of the equinoxes, later discovered by Hipparchos).

vii   The dates of approximate conjunctions of the Moon and of each of the star-like planets with the 31 reference stars and the distance of the body above or below the star given in 'cubits' of 2° and 'fingers' of 5' of arc.

viii   Various pieces of meteorological information, such as rain heavy enough to require the removal of sandals or light enough not to require this action, storms and cloud conditions.

ix   The quantity of certain staples, for example barley, dates, sesame or wool, which could be bought for a shekel (a weight) of silver.

x   Notable events during the month; the diaries record Alexander's entry into Babylon and his later death there.

The early diaries were no doubt important in the generation of the goal texts but later diaries were crucial in the generation of the almanacs and ephemerides which were the products of the late fourth, the third and the second century BC when Babylonian astronomy reached its zenith.

The second factor leading to a scientific status was the early search for regularities, for example the eclipse cycles and the resonance periods of the goal texts; later, say from about 300 BC onwards, quite sophisticated mathematical techniques were used to generate from a modicum of observational data very precise and usually very accurate ephemerides, tables like the modern almanacs giving accurate positions at specified times, for the Sun, the Moon and the five star-like planets. The phenomena calculated in these ephemerides included the dates and zodiacal positions of conjunctions, disappearances and reappearances before and after conjunction, and the oppositions of the three superior planets. In the case of the Sun and the Moon, probable eclipses were also included.

The mathematics used in the later Babylonian period had been developed in the first half of the second millennium but not put to astronomical use at that time. This mathematical system was predominantly arithmetical but it did have some incipient algebraic qualities (see Neugebauer 1969). It was predominantly practical in that the problems it addressed were mensurational (areas of various plane figures and volumes of various solids), proportions (fractional shares in an estate or parts of income to be paid in taxes), and so on. There were also clearly pure mathematical concerns. Amongst these were the generation of increasing and decreasing arithmetic and geometric progressions; this exercise proved to be of great value in late Babylonian mathematical astronomy. Another example of a pure mathematical concern was the generation of Pythagorean triples, $a^2 + b^2 = c^2$, where a, b and c are the sides of a right-angled triangle – this about a thousand years before Pythagoras. 'If the base of a stick c cubits long is pulled out from the bottom of a wall by a cubits, how many cubits (b) up the wall will its top be?' Such triples could be generated by the Babylonians because they had mastered quadratic equations.

The Babylonian number system was sexagesimal and positional. The first integer or 'digits' position provided for numbers between 1 and 59 in our system; the second or 'sixties' position to the left of the first position provided for numbers between 60 and 3599; the third position for numbers between 3600 and 215 999; and so on. Below the first integers or 'digits' and written to the right were a succession of fractions: first, the sixtieths from $\frac{59}{60}$ to $\frac{1}{60}$ in descending value; second, the 3600ths from $\frac{59}{3600}$ to $\frac{1}{3600}$; and so on.

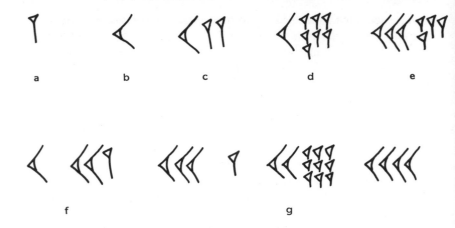

Fig. 3.1   *Babylonian numerals. a is 1, b is 10, c is 12, e is 34, f is (10 × 60) + 21 = 81, and g, if we assume the last two values to be fractions, is (30 × 60) + 1 + ($\frac{29}{60}$) + ($\frac{40}{3600}$) = 1801 + $\frac{1780}{3600}$.*

Two symbols were used to record these numerals: an upright wedge for 1 and repeated for up to 9, and a rotated 'vee' for 10 (see Fig. 3.1). There were two sources of ambiguity in this method of recording numbers. First, there was no analogue to our decimal point so that a Babylonian reader had to resort to his knowledge of the context in which the numeral was reported to decide what was an integer and what a fraction; for example, if the position of a planet within a sign of the zodiac is given as 23, 46, 7 *lu*, then this is to be read as 23°46'7" in *Aries*, though it could also be 0°23'46"7''' in *Aries*; but a clue would be given by where the phenomenon occurred in the preceding sign, *Pisces*. Second, until quite late no symbol was used to indicate that a place was unfilled; it is said that the symbol for nought is an Indian invention but it is almost certainly a late Babylonian invention. Much later the Mayas who used a vigesimal positional number system introduced a symbol for nought. The need for such a symbol seems to have come to be recognized in all positional number systems whether decimal, vigesimal or sexagesimal.

Babylonian calculations were conducted by means of tables. First, there were multiplication tables in which numbers up to 20 were multiplied by 2, 3, 4, ... 60; beyond 20, the numbers 30, 40, 50 and 60 were so multiplied and some few numbers greater than 60 were so multiplied. For a given product, say 5×10, the table yielded an immediate answer; for other products, say 45×10, one needed to look up 40×10 and 5×10 and add the two products together.

Division was conducted by means of tables of reciprocals. In a sexagesimal system, the reciprocal of 2 is $\frac{30}{60}$, of 3 is $\frac{20}{60}$, of 4 is $\frac{15}{60}$ and so on. To divide, say, 72 by 4, one multiplies the number 72 by the reciprocal of 4, which is found by consulting a table of reciprocals to be $\frac{15}{60}$. The product of 72 and $\frac{15}{60}$ is found in the multiplication table to be 18.

These old Babylonian mathematicians were uneasy with irrational numbers. Thus the sexagesimal reciprocals of 7, 11, 13, 14, 18 and 19 are irrational or non-terminating numbers and so did not appear in the tables. Later some irrational numbers were approximated, thus the value we call pi became $3 + \frac{7}{60} + \frac{30}{3600}$ (3.125 in decimal notation instead of 3.1416 . . .) and the square root of 2 became $1 + \frac{24}{60} + \frac{51}{3600} + \frac{10}{216000}$ (or 1.4142129 in decimal notation instead of 1.4142135 . . .). For a fuller account of old Babylonian mathematics Neugebauer (1969) should be consulted.

In the period from 500 BC to the end of the pre-Christian era, various astronomical texts were produced. Some were improved versions of similar texts going back to, say, 750 BC, such as the diaries already mentioned. Some were more sophisticated or better worked out versions of the results of earlier conceptions, for example the goal year texts. About 300 ephemerides, akin to modern astronomical ephemerides, and a smaller number of two types of almanacs which are special kinds of ephemerides have been found.

In their earlier forms the diaries may have been begun in about 747 BC if we accept a clue from Ptolemy, who said he had records of eclipses from the reign of Nabonasser who ascended the throne in 747 BC. There may have been earlier records not available to Ptolemy. On the other hand the earliest eclipse tables are almost certainly in part predictions of possible eclipses rather than observations of eclipses; indeed we have already seen that they could not all have been records of eclipses observed from Babylon. From about 400 BC the diaries settled down into the final form described above.

The earlier instructions, pre-400 BC in the goal texts, for the phenomena of Venus 'go back to the record [presumably a diary] for eight years before and subtract four days', 'for the phenomena of Mars go back to the record for 47 years before', and so on, emerged in goal texts in which for a given year the various planetary phenomena – reappearances, disappearances, first and second stations and in the case of Mars, Jupiter and Saturn, oppositions – were given dates in terms of the resonance periods mentioned. There is a larger number of later goal texts. They represent a systematic mode of generating predictive values based on the earlier stated resonance periods. Later goal texts designated a year and then reproduced for the planetary phenomena of an earlier year the dates and zodiacal positions of reappearances, first stations, oppositions (in the case of Mars, Jupiter and Saturn), second stations and disappearances.

From about 500 BC the Babylonian astronomers, as reported in Chapter 2, had divided the ecliptic into twelve segments each 30° in longitude and given each the name of the zodiacal constellation covering much the same segment. The longitude of the Sun, of the Moon or of one of the five star-like planets was expressed as so many degrees, minutes and seconds of arc (indeed sometimes 'thirds' or sixtieths of a second of arc) within the sign. This practice was continued by Ptolemy and by medieval Arabian and late medieval European astronomers. In addition the latitude of the Moon (in relation to the ecliptic) was also stated in units which were $\frac{1}{72}$ of a degree, which is 50″ of arc. There seems to have been little if any concern in Babylonia with the ecliptic latitudes of the five star-like planets as distinct from their north-south separation from the reference stars.

The system of twelve zodiacal signs each of 30° in longitudinal extent provided a superior reference framework for locating planetary positions to that provided by the system of 31 reference stars. The latter was possibly earlier but we have no clear evidence of that. The zodiacal signs were of equal length whereas the spacing of successive pairs of reference stars varied from 0°3′3″.6 (between *delta Scorpionis* and *beta Scorpionis*) to 63°24′ (between *delta Capricorni* and *eta Piscium*). Further, latitudes were assessed in respect of the ecliptic from a fixed reference plane whereas in the system of reference stars they were assessed from the nearest reference star, which might be up to 9°54′ north of the ecliptic or up to 7°0′ south in 301 BC.

The superiority of the zodiacal reference system in practice depended upon some observational aid(s) to assist in determining angular separations. We have no clear evidence of what the Babylonians used but whatever they were they provided amazing precision.

In addition to the diaries and the goal texts which have already been described, there remain from the last three or four centuries before the beginning of the Christian era three other main classes of astronomical texts. They are (a) ephemerides, (b) almanacs, and (c) normal star almanacs. The texts that we have come from Babylon or from Uruk. No doubt there were other centres which produced similar texts but their products have not so far been discovered or identified (see Sachs 1948).

A text giving an ephemeris, of which there are about 300 in various states of completeness, ordinarily covers a year or a run of years, sometimes day by day but more usually from one occurrence of some phenomenon to the next. An ephemeris gives one or more of the following sorts of information.

1.   In respect of the Moon:

a   Whether the calendar month just ended was 'full' (30 days) or 'hollow' (29 days) and the time between sunset and the setting of the first visible crescent whose appearance marked the beginning of the month.

b   Dates of lunar eclipses, those observed or those predicted but occurring below the horizon or obscured by cloud, and dates of solar eclipses, predicted to be possibly visible or not possibly visible from the specified site; the time of lunar opposition to and conjunction with the Sun; the longitude in ecliptic or zodiacal terms of the eclipsed or possibly eclipsed body was stated and in the case of possible lunar eclipses, an eclipse magnitude (essentially a lunar ecliptic latitude).

c   For several dates around the middle of the month when the moon was full or nearly so, times between (i) moonrise before sunset and sunset itself, (ii) sunset and moonrise after sunset, (iii) moonset before sunrise and sunrise, and (iv) sunrise and moonset after sunrise.

d   The date when the Moon is visible for the last time before sunrise and the interval of time on that occasion between moonrise and sunrise.

2. In respect of the five star-like planets:

a   For Mercury, the dates in months and days and ecliptic longitudes in degrees, minutes, seconds and sometimes thirds from the beginning of a sign of the zodiac of the following phenomena: (i) the first visibility of the planet after sunset following a short period of invisibility around the planet's inferior conjunction with the Sun, (ii) the last visibility of the planet before sunrise before the planet's inferior conjunction, (iii) the first visibility after sunset after superior conjunction, and (iv) the last visibility before sunrise before superior conjunction.

b   For Venus, there were added to the dates and the zodiacal position of the two pairs of phenomena corresponding to those mentioned for Mercury, the dates and zodiacal positions of the first and second stations, that is the beginning and the end of the phase of apparent retrograde motion.

c   For Mars, Jupiter and Saturn (which have only the equivalent of a superior conjunction, the inferior conjunction being replaced by an opposition), the dates and zodiacal positions of (i) the first visibility before sunrise after conjunction, (ii) the first station at the beginning of the retrograde phase, (iii) opposition, (iv) the second station, and (v) the last visibility after sunset before conjunction.

d   The ephemerides often provided information on the daily apparent motion of the planets. Such apparent motion had a maximum positive (eastward) value when Mercury and Venus were around superior conjunction and when Mars, Jupiter and Saturn were around conjunction.

Some examples of planetary and lunar ephemerides will be discussed in some detail later. They are clearly an extension of the earlier diaries.

The almanacs usually cover a single year and provide a section for each month. The special group of almanacs called normal star almanacs report positions of the Moon and of the five star-like planets relative to the 31 reference stars as well as to the signs of the zodiac. The ordinary almanacs usually gave the following information:

1.   Three data relevant to the Moon, namely, the date of first visible crescent in terms of the length of the preceding month, that

is either 29 or 30 days; the interval between sunrise and moonset immediately after sunrise; and the date of the last visible crescent.

2. Dates and positions in the signs of such planetary phenomena as disappearances before and reappearances after conjunction with the Sun, and in respect of Mars, Jupiter and Saturn first and second stations and opposition.

3. Dates of planetary entry into signs and whether the motion was direct or retrograde.

4. Planetary positions in zodiacal terms at the beginning of the month.

5. Dates of solstices and equinoxes (these were certainly schematic and not real dates of solstices and equinoxes as they are evenly spaced).

6. Date and often time in the day of eclipses, whether observed of not.

7. Often, last visible evening setting of Sirius, first visible morning rising.

The normal star almanacs contained some of this information, omitted some of it and added some. Thus,

1. Some of them added to the interval from sunrise to moonset immediately after sunrise, the intervals from moonset before sunrise to sunrise, from moonrise before sunset to sunset and from sunset to moonrise after sunset.

2. Data as in 2 in the ordinary almanacs were included.

3. Data as in 3 in the ordinary almanacs omitted and position of the planets relative to reference stars substituted.

4. Data as in 4 above omitted.

5, 6 and 7. The data given in the analogous items in the ordinary almanacs.

Why the almanacs were compiled is not clear, nor is the manner of their compilation: they probably had some basis in the diaries and ephemerides.

Table 3.1 Part of a table giving first stations of Jupiter from
SE 113

Year in SE	Position	Motion	Time interval
113	8;6 Cap.	–	–
114	14;6 Aqr.	36°	13m 24d
115	20;6 Psc.	36°	13m 24d
116	26;6 Ari.	36°	13m 24d
117	2;6 Gem.	36°	13m 24d
118	5;55 Cnc.	33°49′	13m 23d
119	5;55 Leo	30°	13m 21d
120	5;55 Vir.	30°	13m 21d
121	5;55 Lib.	30°	13m 21d
122	5;55 Sco.	30°	13m 21d
123	7;6 Sgr.	31°11′	13m 22d
125	13;6 Cap.	36°	13m 24d
126	19;6 Aqr.	36°	13m 24d
127	25;6 Psc.	36°	13m 24d
128	1;6 Tau.	36°	13m 24d
129	7;6 Gem.	36°	13m 24d
130	10;5 Cnc.	32°59′	13m 23d
131	10;5 Leo	30°	13m 21d
132	10;5 Vir.	30°	13m 21d
133	10;5 Lib.	30°	13m 21d
134	10;5 Sco.	30°	13m 21d
135	12;6 Sgr.	32°1′	13m 22d

For each year the position of the first station within a sign is given,
followed by the implied motion from the preceding first station and the
elapsed time.

The genuinely scientific compilations are the ephemerides, to
which I shall now turn. All of the 'wanderers' have variable
apparent velocity (as a result of the excentricity of planetary and
lunar orbits – a fact unknown to the Babylonians).

The Sun's variation in apparent velocity we know to be the
consequence of the Earth's slightly excentric orbit with maximum
angular velocity at perihelion and minimum angular velocity at
aphelion. The Earth's variation in angular velocity is added to the
changes in velocity of the Moon and of the five star-like planets

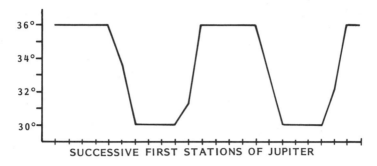

SUCCESSIVE FIRST STATIONS OF JUPITER

Fig. 3.2   *Graph of successive first stations of Jupiter illustrating a step function with a maximum of 36° and a minimum of 30°.*

resulting from their several excentricities. In the cases of the five star-like planets, the apparent slow-down in motion is so great as to appear to be negative, that is retrograde motion, though this is the result primarily of the motion of the Earth, from which their positions are seen.

A modern astronomer would represent these changes in apparent velocity by some form of sinusoidal function depicting a smooth transition from maximum to minimum and back to maximum velocity. The Babylonians used two types of function with jumps or breaks in them. One was a step function and the other a zigzag function. The former may be illustrated by a table giving the dates and positions in the zodiac of the first stations of Jupiter from 113 to 173 in the Seleucid era, i.e. 199 to 138 BC. Extracts from this table are given in Table 3.1 for the first 22 Babylonian calendar years. Column 1 gives the years in the Seleucid era (SE) in which the phenomena occurred; column 2 gives the positions of Jupiter within the zodiacal signs of the successive first stations; column 3 gives the implied successive motions through the zodiac; and column 4 gives the elapsed time in months and days between the successive phenomena. It will be noticed that there is a series involving 36° motion and 13 month 24 day intervals followed by a transitional pair of values which in turn are followed by a series of 30° of motion and a period of 13 months and 21 days with transitional values in between the main series. When graphed as in Fig. 3.2, these values constitute a step function.

Table 3.2   Dates and positions of the second stations of Jupiter, SE 190 to 209

	1	2	3	4	5
SE year	Month, day	Interval	Longitude	Implied motion	Difference in implied motion
190	XII 11		21°49′ Cnc.		
★ 191	XIIa 22	13m 11d	21°20′ Leo	29°41′	−1°3′
193	II 4	13m 12d	20°8′ Vir.	28°38′	+1°48′
★(194)	III 16	13m 12d	20°34′ Lib.	30°16′	+1°48′
(195)	(IV) 1	13m 15d	22°48′ Sco.	32°12′	+1°48′
196	V 17	13m 11d	26°50′ Sgr.	34°2′	+1°48′
★ 197	VII 5	13m 18d	2°40′ Aqr.	35°50′	+1°48′
198	VII 25	13m 20d	10°18′ Psc.	37°38′	−1°0′
★ 199	IX 13	13m 18d	16°56′ Ari.	36°38′	−1°48′
200	IX 30	13m 17d	21°46′ Tau.	34°50′	−1°48′
201	XI 15	13m 15d	24°48′ Gem.	33°2′	−1°48′
★ 202	XII 28	13m 13d	26°2′ Cnc.	31°14′	−1°48′
204	I 10	13m 12d	25°28′ Leo	29°26′	−1°33′
★ 205	II 21	13m 11d	24°21′ Vir.	28°53′	+1°48′
206	III 4	13m 13d	25°2′ Lib.	30°41′	+1°48′
207	IV 18	13m 14d	27°31′ Sco.	32°19′	+1°48′
★★208	VI 4	13m 11d	1°48′ Cap.	34°17′	+1°48′
209	VI 23	13m 19d	7°53′ Aqr.	36°5′	

Years marked with one asterisk have a second twelfth month, those with a double asterisk a second sixth month.

There are also Jupiter ephemerides using four steps, namely 30°, 33°, 35° and 36°, and Mars ephemerides using six steps. The Mars ephemeris embraces two signs in each step. Each step is divided into a certain number of intervals, for example 12 intervals for *Capricornus* and *Aquarius*, 16 for *Pisces* and *Aries*, 24 for *Taurus* and *Gemini*, 36 for *Cancer* and *Leo*, and so on. The period between successive occurrences of, say, the first station of Mars is eighteen intervals which cover two and three-quarter signs in the neighbourhood of *Sagittarius*, *Aquarius* and *Pisces*, but little more than a sign in the neighbourhood of *Cancer* and *Leo*. This more frequent and more marked variation in the step function for Mars

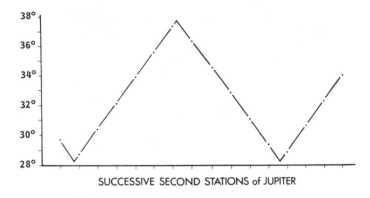

Fig. 3.3   *Graph of successive second stations of Jupiter*
illustrating a zigzag function with a maximum of 37°45' and
a minimum of 28°15'.

accommodates the greater variation in the apparent velocity of
Mars as a result of its substantial orbital excentricity.

The zigzag function may be illustrated by means of an
ephemeris for the second station of Jupiter for nineteen years from
190 to 209 in the Seleucid era (about 121 to 102 BC) (see Table
3.2). Column 1 gives the dates in years in the Seleucid era and
months and days within the years; column 2 gives the intervals
between successive second stations in months and days; column 3
gives the zodiacal positions of the successive second stations; and
column 4 gives the implied motion from the preceding second
station, thus the entry 29°41' is obtained by subtracting 21°49'
Cancer from 21°30' *Leo*. When we plot in Fig. 3.3 the zodiacal
position of the second stations of Jupiter as given in Table 3.2, we
obtain a zigzag function increasing by 1°48' per second station
from about 28°15' to about 37°45'. This is a still closer approximation than any of the step functions to the sinusoidal function.

In the ephemerides generated by means of step functions and
zigzag functions some empirical bases obviously were used. The
angular difference between the maximum and the minimum
values and the time intervals between maxima clearly approximated the empirical values. Nevertheless there appears to have
been some rounding of values at key points. On the other hand
the calculated positional values for specified phenomena closely

Table 3.3   Part of a luni-solar ephemeris from month XII of year 207 in the Seleucid era

0	1	2		3	4	5		6	7		15
XII	29,8,39,18	2,2,6,20	Ari.	2,56	1,32	6,5,30	sig	11,30	3,59,52,30	30	15,40
208 I	28,50,39,18	52,45,38	Tau.	3,14	1,23	9,46,30	sig	11,16,10	4,22,22,30	30	17,30
II	28,32,39,18	29,25,24,58	Tau.	3,26	1,17	5,54	sig	11,52,10	4,14,1,40	29	13
III	28,14,39,18	27,40,4,14	Gem.	3,34	1,13	2,1,30	sig	12,28,10	3,51,31,40	30	–,40
IV	28,24,40,2	26,4,44,10	Cnc.	3,32	1,14	1,51	bar	13,4,10	3,29,1,40	29	–, –
V	28,42,40,2	24,47,24,18	Leo	3,24	1,18	2,43,30	nim	13,40,10	3,6,31,40	30	17, –
VI	29, ,40,2	23,48,4,20	Vir.	3,9	1,25	6,36	nim	14,16,10	2,44,1,40	29	12
VIa	29,18,40,2	23,6,44,22	Lib.	2,51	1,34	9,16	nim	14,52,10	2,21,31,40	30	8,40
VII	29,36,40,2	22,43,24,24	Sco.	2,36	1,42	5,23,30	nim	15,4	1,59,1,40	30	9,10
VIII	29,54,40,2	22,38,4,26	Sgr.	2,27	1,46	1,31	nim	14,28	2,8,36,30	30	16,30
IX	29,51,17,58	22,29,22,24	Cap.	2,27	1,46	2,21,30	bar	13,52	2,31,7,30	29	10 us
X	29,33,17,58	22,2,40,22	Aqr.	2,36	1,42	3,14	sig	13,16	2,53,37,30	30	15,10
XI	29,15,17,58	21,17,58,20	Psc.	2,50	1,35	7,6,30	sig	12,40	3,16,7,20	30	19,20

Columns 0–7 are explained in the text. Column 15 gives the number of days in the preceding month and the time from sunset to moonset in time degrees (15° equals one hour). Some of the values given are restorations.

approximate those that we can calculate back to those times from modern considerations. This claim is well illustrated by a luni-solar ephemeris for the years 208 to 210 in the Seleucid era (approximately 104 to 107 BC). Thirteen lines of its 39 lines and eight of its seventeen columns as reported by Aaboe (1974) are set out in Table 3.3.

Column 0, a restoration of a broken-away segment of the tablet, gives years in the Seleucid era (SE) and months in the Babylonian calendar: the second to the thirteenth line cover twelve months in 208 SE. Column 1 gives the progress of the Sun in degrees, minutes, etc., since the preceding date. As the phenomena so dated are lunar-solar conjunctions, column 1 also states lunar progress plus 180°. The value in column 1 when added to the preceding value in column 2 gives the next position of the conjunction. Thus 0;52,45,38 (*Taurus*) the position of the conjunction of the Sun and Moon in month I of 208 SE plus 28;32,39,18, the advance of the Sun in the following synodic month, yields 29;25,24,56 (*Taurus*) as the zodiacal position of the next conjunction in month II. Values in column 1 are generated by a zigzag function having a maximum of 30;1,59,0 and a minimum of 28;10,39,40 and a period of one synodic month. Columns 3 and 4 give the length of sunlight (sunrise to sunset) and the length of half the 'night' (sunset to sunrise) in time degrees and minutes where 15° equals one modern hour. These values have been generated by means of zigzag functions with a period of one year. Column 5 gives the latitude of the Moon. These values have been generated by a zigzag function whose period is the draconic month, 27.2212 days. The words 'sig' and 'nim' indicate a negative and a positive latitude respectively and 'bar' indicates a possible solar eclipse. Column 6 gives the lunar velocity at conjunction in degrees, minutes and seconds per day. Its values are generated by a zigzag function whose period is the anomalistic month. Column 7 gives the excess in time degrees over 29 days of the interval between one conjunction and the next generated by a zigzag function with a period of the anomalistic month. These values imply that the mean synodic month is 29;31,50,8,20 days or in decimal notation 29.5305941 days which is remarkably close to our best value, 29.5305882 calculated on modern considerations for 100 BC. Further columns in the original table correct the intervals between conjunctions for variations in solar velocity and give the time of the conjunctions in time degrees after midnight, and end by stating the intervals between successive first visible crescents and the intervals on those occasions between sunset and moonset.

When the times of conjunctions reported in this ephemeris are compared with times calculated from modern considerations they are found to have a mean constant error of +1.5 hours with a variable error of 1.0 hour, which is a truly remarkable achievement for the first century BC. As stated above it would seem that through a series of observations over several cycles, the late Babylonian astronomers were able to establish good approximations to the maxima and the minima of the changes of zodiacal positions amongst such phenomena, to the intervals between these maxima or minima and the approximate absolute positions in time (expressed in calendar years, months, days and hours) of these events. The diaries give us some indication of the data on which these parameters of the ephemerides were based but there are many remaining uncertainties.

Late Babylonian astronomy contributed several important constituents to Greek astronomy. First, it bequeathed a number of planetary (in the Greek sense) periods, most of them highly accurate. Amongst these were the synodic periods of the Moon and of the five star-like planets. In addition, the periods of draconic and anomalistic months were distinguished from the synodic month. The Babylonians, however, had to leave to the Greeks the distinction between the sidereal and tropical years, a distinction later to be explained. Second, it bequeathed a great deal of data. The Greek astronomers often mistook Babylonian predictions for observations. These predictions, however, were often so good, as we have seen, that they could serve well if treated as observational data. Third, the Greek astronomers, from at least Hipparchos onwards, adopted the Babylonian conventions of dividing the circle into 360 degrees, of dividing the ecliptic into twelve signs each of 30 degrees, and of dividing a degree into 60 minutes, a minute into 60 seconds and often a second into 60 'thirds'. Fourth, Greek astronomers adopted and usually improved the Egyptian and Babylonian 'sun-clocks' and 'water-clocks', primarily for the measurement of the passage of time. By contrast, whereas the Greeks adopted the products of the Babylonian mathematical methods, as reported above, they substituted geometrical analyses for the Babylonian arithmetico-algebraic analyses, and developed mechanical models of apparent planetary motion. Apart from quite late Babylonian thought and then under Hellenistic influence (cf. Seleukos, to be mentioned later) Babylonian astronomy lacked the physical or quasiphysical models based on mathematical analysis which we find in Greek astronomy from Eudoxos to Ptolemy. Despite the relatively sophisticated mathematical analyses of Babylonian astronomy the physical conceptions of stellar and planetary motions were completely misguided.

Though we sometimes have difficulty in providing the details of the contributions of many of the early Greek astronomers, we have no difficulty in most cases in providing the broad outlines of the contribution and the name of the contributor. It is quite the reverse with the near contemporary later Babylonians; we have in great detail their contributions such as diaries, goal texts, ephemerides and so on, but scarcely any names. Strabo, the Greek geographer and historian (*circa* 43 BC–AD 23) refers to the Chaldean mathematicians and astronomers Kidenas, Naburiannous and Soudines. Later Vettius Valens and Pliny mention some of these names, which are probably Graecized forms of Akkadian names. On one fairly late tablet we have the inscription 'Computing tablet of Kidinnu for the years 208 to 210'. Kidinnu may have been the astronomer who gathered the information on which the ephemeris was based or a scribe who made the computations according to some rule. There are also references to Naburianna who it has been suggested worked out the length of the year as 365.2609 days. The name Anu-she-shu-idinna appears on another tablet and this may be the origin of Strabo's Soudines. Perhaps the Babylonian priest-astronomer-scribes were less concerned with personal recognition and more concerned with discharging their duty to the gods.

CHAPTER FOUR

EARLY GREEK ASTRONOMY

*W*e have some pre-scientific Greek astronomy in Hesiod's *Works and Days, c.* eighth century BC. He says, for instance, that when the Pleiades make their heliacal rising (in May of our calendar) the farmer should sharpen his sickle to harvest the crop and when they are making their heliacal setting (in our November) he should begin to plough. Again, when 60 days after the winter solstice the star Arcturus rises in the early evening and remains visible for the whole night, the grapes should be pruned.

In the seventh and sixth centuries there was much philosophical speculation on cosmological issues with some astronomical reference, but as we shall see all of this was pre-scientific. Some proto-scientific astronomy began in the fifth century with attempts to establish lowest common multiples of such periods as the day, the month and the year, and to generate mathematical models explaining apparent planetary motions. Genuine, if still not fully developed, scientific astronomy was propounded by Hipparchos in the second century BC and by Ptolemy in the second century AD.

It had earlier been thought that Babylonian astronomy was almost entirely observational and almost completely atheoretical, whereas Greek astronomy was poor in observation, but strong in theory. This involves a serious misunderstanding of both Babylonian and Greek astronomy. As we have seen, only the Babylonian diaries were predominantly observational and even they included a lot of predictions. The eclipse tables, the almanacs and the ephemerides which may have seemed to be records of observation were in fact mainly predictions made on ingenious mathematical bases applied to approximate observational data.

There are many fruitful sources of information about Greek astronomy. Dreyer (1905) is a good source on early Greek pre-scientific, early scientific and established scientific astronomy, as are Lloyd (1970, 1973) and Pannekoek (1961). Dicks (1970) is valuable on the early stages. The *Dictionary of Scientific Biography* (Gillespie 1970–4) is very valuable on the contributions of individual Greek astronomers. Outstanding but not always easy for the general reader to follow is Neugebauer (1975); his 1969 book is less comprehensive but easier to penetrate.

A series of Milesian philosophers, Thales (seventh century BC), Anaximander (*c.* 611 to 545 BC) and Anaximenes (late sixth century BC), speculated about the structure and composition of the universe. We have to rely largely on not always consistent secondary sources. They seem to have thought of the Earth as a disc or in some cases as a domed column floating on water or air and covered by a hemispheric vault. There was disagreement about the relative distances of the celestial bodies. The Moon was usually considered to be nearest but some put the Sun and the star-like planets beyond the fixed stars whereas some put the Sun next after the Moon followed by the star-like planets with the fixed stars being regarded as the most remote. Anaximander had a strange conception of the Sun and of the Moon. He considered them to be wheels which rotated and which had hollow rims filled with fire. Each rim had an aperture through which the fire could be seen. A temporary closure of the aperture resulted in an eclipse. Thales is said by Herodotos to have predicted a solar eclipse during a battle between the Medes and the Lydians in Asia Minor; this was presumably the eclipse of 584 BC. It has been suggested that he may have been using the Babylonian eclipse cycle, but this is implausible. Herodotos said that Thales had predicted the eclipse to within the correct year; had he been using the Babylonian cycle he would have been within the correct month. Further, the

Babylonian eclipse cycle did not enable the prediction of a solar eclipse as being visible from a particular place. These philosophers had various suggestions about the basic substance of which the universe was constituted; water, air and fire were severally nominated.

Pythagoras (*c.* 580–500 BC) and his followers believed that numbers were the basis of all things. They were successful in showing that the musical scale could be resolved in terms of simple ratios and in showing several basic mathematical ratios. They became increasingly mystical and secretive. As a consequence of their keeping their views to themselves few of their doctrines became public. They may have extended their numerical analyses to astronomical matters but we have little information on that. They do seem to have asserted the sphericity of the Earth, although this seems to have been worked out more fully by Parmenides who will be discussed later. A Pythagorean, usually said to be Philolaus (latter half of the fifth century), asserted that there was an unseen central fire around which the Earth revolved in 24 hours, beyond the Earth the Moon revolved in 29.5 days and beyond the Moon the Sun revolved in presumably a year. Between the Earth and the central fire there was assumed to be an invisible counter-earth revolving also in 24 hours and so perpetually obscuring the central fire for observers on the Earth. The placement of the Earth, the Moon and the Sun would explain solar eclipses but without offsetting the lunar and solar orbits would require monthly solar eclipses. Though it was claimed that the theory explained the greater frequency of lunar eclipses, it is difficult to see how it accounted for lunar eclipses at all.

Parmenides (early fifth century), possibly influenced by the Pythagoreans, asserted the sphericity of the Earth and extended the concept to the whole universe, which was assumed to be arranged in a series of concentric layers or spheres surrounding the Earth. It is obscure as to the relation of these layers to the various celestial bodies. Parmenides is also credited with the identification of the morning and evening star as one and the same body. Empedocles (mid-fifth century) assumed that there were two celestial spheres, one of fire and the other of air with a little fire. The circulation of these two spheres in alternating dominance was deemed to result in the succession of day and night.

The early Greek philosophers who paid any attention to

astronomical matters were clearly pre-scientific. From about the fifth century Greek astronomy began to enter a proto-scientific phase, developing into a genuine early scientific phase. Plato (428–347 BC) played an important role in this development. He was not an observer or a contributor to astronomical theory. He repeatedly insisted on the importance of the teaching of astronomy in the education of the various grades of citizens, even the ordinary citizen should know enough to understand the problems of the calendar. He also stressed the need to mathematize astronomy; this may have been a Pythagorean influence. He is said to have formulated the requirement that anomalous or uneven planetary apparent motion should be reduced to component circular uniform motions. The concentric spheres of Eudoxos, of Kallippos and others was one attempt to meet this requirement; the deferents and epicycles of Hipparchos, Ptolemy and others was another. From about this time Greek astronomy began to enter a scientific phase, both in terms of adopting Babylonian values which were becoming available to it and establishing further observational values of its own, and in terms of explanatory theories usually in the form of mathematical, usually geometrical, models.

A number of more or less accurate observational values were established by Greek astronomers from the beginning of the second half of the first millennium BC. An early concern was the attempt to find a relationship between the synodic months and the year. Kleostratos (c. 500 BC) is said to have claimed that 99 months of about 29.5 days or a little more on average equalled eight years of presumably 365.25 days each. If five years had twelve synodic months, the remaining three would need to have thirteen, usually said to be the third, the sixth and the eighth year. This cycle was called the *oktaeteris*. However this 'equation' was calculated, it never came out quite right. Eight years of 365.25 days (the usually accepted value) total 2922 days. Ninety-nine months of 29.5 days total 2920.5 days, whereas if the value 29.53 is used the total is 2923.47 days. In 432 BC Meton enunciated another cycle in which nineteen years (say 6939.75 days) were equated with 235 months, say 6939.55 days if one takes the mean synodic month to be 29.53 days. Meton adopted a total of 6940 days; this value implied that the solar year was 365.26316 days and that the lunar synodic month was 29.53149 days, half an hour and two minutes respectively too long. Kallippos, *c.* 370 BC, improved the Metonic cycle by taking four of Meton's periods but dropping one day from the set of four. Thus 76 years and 940 synodic months totalled 27 750 days, making the year 365.25 days (eleven minutes too long for the tropical year) and the mean synodic month 29.53085 days (22 seconds too long). Ptolemy, second century AD, made use of the Kallippic cycles in his dating of astronomical observation. Hipparchos, second century BC, suggested a further improvement which seems not to have come into practical use. He took four Kallippic cycles and dropped a further day, equating 304 years and 3760 synodic months with 111 035 days. This implied the year to be 365.25 less about $\frac{1}{300}$ days or 365.2467 days (about 6.6 minutes too long) and the mean synodic month to be 29.530585 days (less than one second in error).

The cycle which was enunciated by Meton in 432 BC may have been a Babylonian discovery. We find it applied to the Babylonian calendar in exact form from *c.* 380 BC and in approximate form from *c.* 480 BC (see O'Neil 1975)

Euktemon, a contemporary of Meton, is said to have established that the intervals between the equinoxes and solstices were unequal. He differed here from the Babylonian astronomers who used schematic evenly spaced dates. Beginning with the spring equinox, Euktemon is said to have found the intervals to be 93, 90, 90 and 92 days respectively. It has been doubted, however, that he discovered more than that the period from the summer solstice to the winter solstice was a few days shorter than the period from the winter solstice to the summer solstice. While he would have been able to approximate the dates of the solstices within a day or two, it is unlikely for reasons given below that he could have found the dates of the equinoxes to within several days. In the fourth century, Kallippos, who may have been able to date the equinoxes, is said to have improved on Euktemon's values by proposing 94, 92, 89 and 90 days for the four seasons. Modern calculation indicates that in Kallippos's times the values were 94.1, 92.3, 88.6 and 90.4 days.

The inequality of the astronomical seasons is the result of the change in apparent velocity of the Sun, now greatest in early January (when the Earth is at perihelion) and least in early July (when the Earth is at aphelion). The interval between successive perihelia, the anomalistic year, is 365.25964 days which is 0.01845 days longer than the tropical year on which our calendar is now approximately based. After 200 calendar years the date of perihelion falls almost 35 days later than it did. This affects over a long period of time the relative lengths of the four astronomical seasons. Thus at the time Kallippos made his estimates perihelion would have occurred in mid- to late October by our calendar, thus shortening his autumn as compared with ours.

In principle the dates of the equinoxes should be more readily determinable than the dates of the solstices. At the equinoxes the Sun changes $0°.5$ in declination (separation from the celestial equator) in a little over 30 hours, whereas nearing the winter solstice it changes $0°.5$ in declination in a little over eleven days and nearing the summer solstice in about twelve days. At the summer solstice the Sun rises furthest north and has its highest altitude on crossing the meridian. The changes around the crucial date are so small that about a week is required to produce an appreciable naked-eye difference. Thus a bracketing operation extending over about two weeks would enable the dating of the solstice to within a day or so. With the aid of an accurate compass enabling one to establish due east and due west the date of the equinoxes can be established to about a day, for on those occasions the Sun rises near due east and sets near due west. In ancient times, however, the best compass was the Sun itself and hence of little value for the exercise indicated. The magnetic compass seems not to have been available in the west until about the end of the first millennium AD. Hipparchos almost certainly had the use of an equatorial armilla, a ring fixed in the equatorial plane. If an equinox occurred between sunrise and sunset the part of the ring nearest the Sun cast a shadow on the part furthest from the Sun. Apart from the effects of refraction when the Sun was near the horizon, the equatorial armilla enabled the determination of the time and date of an appropriate equinox to a matter of hours or less. Whether Kallippos had such an instrument enabling him to improve on Euktemon's seasonal values is not known.

Some other early Greek observations are worthy of mention either because of the methods used or because of the good approximations of the values established. Aristarchos, a native of Samos but possibly working in Alexandria at the time (third century BC), recognizing that the Moon is seen in reflected light from the Sun, argued that when the Moon was seen as a half-disc

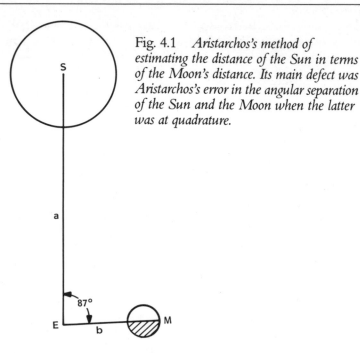

Fig. 4.1 *Aristarchos's method of estimating the distance of the Sun in terms of the Moon's distance. Its main defect was Aristarchos's error in the angular separation of the Sun and the Moon when the latter was at quadrature.*

it must lie at the right-angled corner of a triangle formed by Earth, Moon and Sun (see Fig. 4.1). He took the angle Moon-Earth-Sun to be 87°, a gross underestimate, and deduced that the Sun is about eighteen to twenty times more distant than the Moon and that its diameter is six to seven times greater than that of Earth (he used here the diameter of the Earth's shadow cone relative to that of the Moon as revealed at total lunar eclipses). The method is ingenious, but it led to grossly erroneous results through the estimation of the angle Moon-Earth-Sun as 87°. It is slightly greater than 89°50′. Using the higher angular value, the Sun would be inferred to be almost 390 times more distant than the Moon and its diameter to be almost 114 times the diameter of the Earth.

Eratosthenes, third century BC, believing the Earth to be a sphere, conceived an ingenious way of estimating its circumference. At the summer solstice, he had an assistant stationed in Syene in upper Egypt, which he believed was located on the tropic of Cancer, observe when the Sun was directly overhead at midday. There are several accounts of how the assistant did this, for example from a well when the Sun was seen to be directly

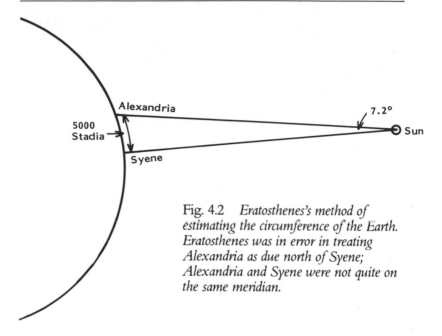

Fig. 4.2 *Eratosthenes's method of estimating the circumference of the Earth. Eratosthenes was in error in treating Alexandria as due north of Syene; Alexandria and Syene were not quite on the same meridian.*

overhead or from a gnomon casting no shadow. Eratosthenes at Alexandria used a gnomon casting a shadow in a hemispherical sundial and estimated the angle of the Sun on the summer solstice (see Fig. 4.2). Using this angle, less the supposed 90° at Syene, and the roughly measured distance on a direct line from Syene to Alexandria, Eratosthenes concluded that the Earth had a circumference of about 250 000 stadia. In those days three different linear measures were called stadia. It is not clear which one Eratosthenes was using. If we select the most favourable, his inference about the circumference of the Earth is very good.

Eratosthenes also tried to estimate the obliquity of the ecliptic to the celestial equator. The Babylonians seemed to have recognized that the ecliptic, their main reference base, zigzagged across the equator, but they seem not to have been concerned about the angle of the ecliptic to the equator. Eratosthenes by measuring the angular separation of the Sun at the two solstices assessed the obliquity of the ecliptic to be between 23°51′ and 24°. Subsequent opinion favoured the lower value.

The early Greeks interested in astronomy did a good deal of speculating about the order of distance of the *planetai* and about the reasons for their apparent behaviour. On the basis of sidereal periods as seen from the Earth, it seemed that the Moon (27.3 days) was nearest, then Mercury, Venus and the Sun (all about 365.25 days), followed by Mars (almost two years), Jupiter (almost twelve years) and Saturn (almost thirty years). That Mercury and Venus had different synodic periods (the modern values are 116 days and 584 days) suggested that Mercury was next after the Moon, then Venus and then the Sun, but there were other views about the order of distance of these three bodies. The conclusions being drawn here assume that the seven *planetai*, including the Sun, and the fixed stars revolve around the Earth, a frequent assumption not only amongst the Greeks, as it seems to fit the evidence of our senses.

Some early Greeks, however, did not make this geocentric assumption. Perhaps Herakleides, fourth century BC, did not. His own writings have not survived and later authors who cite him, though agreed that he suggested the rotation of the Earth as the explanation of the diurnal apparent motion of all the celestial bodies, are not all clear on whether he asserted the revolution of the Earth in a circle around the Sun. Aristarchos, third century BC, is reported by Archimedes to have propounded the hypotheses that the Earth both rotates around its polar axis and revolves around a stationary Sun. The same propositions were put forward about a century later by Seleukos, probably a Greek resident in Mesopotamia but possibly a Babylonian who had assumed a Greek name. It was sometimes said that while Aristarchos advanced these propositions as hypotheses, Seleukos advanced them as truths (see Dreyer 1905). Whether the distinction was correctly applied to Aristarchos and Seleukos or not, the distinction is important. One might make analyses or assumptions of a simple sort from which more complex observed events might be deduced without giving the analyses or assumptions other than an instrumental (or 'convenient fiction') status. On the other hand one might assert that only realities, believed states of affairs in nature, should be so used. This is an important issue in dealing with early and late Greek astronomical theorizing.

If we are travelling in a boat being carried downstream in a river, the banks appear to be travelling backwards. It is not always easy to determine from purely observational data whether the boat is going forwards or the banks are going backwards. In some circumstances it may be more convenient to assume that the latter is happening.

Plato is said to have asked that astronomers reduce observed celestial motions to uniform (in respect of angular velocity) circular motions, but it may have been Pythagoras who made the demand. This is an easy task in respect of the fixed stars. They may be regarded as being fixed to the surface of a distant celestial sphere rotating east to west in about a day with the Earth at its centre.

Eudoxos, fourth century BC, proposed a scheme involving sets of concentric spheres for each of the seven *planetai*. Thus, for the Sun, an outer sphere was deemed to rotate uniformly around a polar axis in about a day and an inner sphere rotated uniformly on an axis borne by the first sphere but on an axis offset from the polar axis by about 24°, and in a year, with the Sun on the equator of the inner sphere, this will result in the Sun tracing the ecliptic. This model did not explain the changes of solar angular velocity which had been discovered by Euktemon before Eudoxos formulated his hypothesis. Eudoxos did a quite strange thing in addition; he introduced a third sphere for the Sun in order to account for a slight (undetectable by any means available to him and in fact non-existent) departure of the Sun from the ecliptic.

We are handicapped in understanding Eudoxos's hypotheses, as his own account has been lost. We have a rough outline given by Aristotle and a much later more detailed account by Simplicius commenting on Aristotle. Simplicius seems to have had many of the details wrong. We are indebted to Schiaparelli who pointed out errors in Simplicius's account and suggested in 1875 a correction of it (see Dreyer 1905). Unfortunately in correcting quantities given by Simplicius, Schiaparelli substitutes values known to him but almost certainly not known to Eudoxos. Nevertheless, I shall follow Schiaparelli's reconstruction, as outlined by Dreyer. The Moon was located on the equator of an inner sphere which rotated in 27.212 days (the draconic month) west to east on an axis offset by 5° from that of the second sphere which rotated in 223 synodic months east to west on an axis offset by about 24° to the polar axis of the outer sphere rotating east to west in about a day (see Fig. 4.3). This scheme accounts for the diurnal east to west motion of the Moon, its monthly west to east motion, its variations in latitude between about ± 5° (from the ecliptic) and the retrogression of the lunar nodes (the points where the Moon

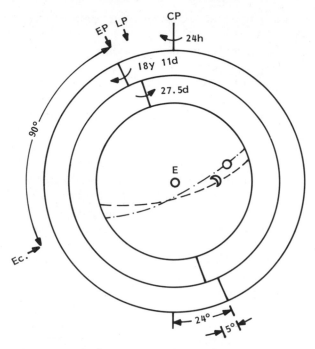

Fig. 4.3   *The homocentric spheres of Eudoxos for the Sun (open circle) and for the Moon (crescent). CP is the North Celestial Pole, EP the North Ecliptic Pole and LP the north pole of the lunar orbit. Other details are set out in the text.*

crosses the ecliptic) in a little over eighteen years. It does not, however, account for the variation in the angular velocity of the Moon within a month; this is much greater than the variation in angular velocity of the Sun within a year.

The schemes for the five star-like planets involved in each case four spheres. The outer one like those for the Sun and for the Moon rotated east to west on a polar axis in about a day, bearing the axis (offset by 24°) of the second sphere which rotated west to east in the zodiacal period of the planet concerned. Simplicius states these to be one year for Mercury and for Venus, two years for Mars, twelve years for Jupiter and 30 years for Saturn. The axis of the third sphere was borne by the second sphere and was located on the ecliptic (the equator of the second sphere). The location of these axes was the same for Mercury and for Venus but different

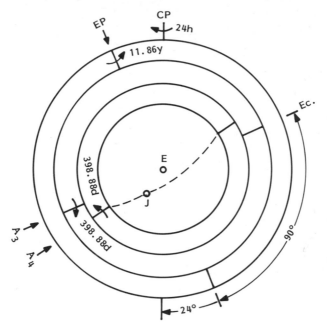

Fig. 4.4   The four homocentric spheres of Eudoxos for the
five star-like planets. The data on angles and periods set out
here are for Jupiter.

for the other three. The third sphere rotated in the synodic period
of the planet concerned; the periods given by Simplicius are 110
days for Mercury (a more exact value would be 116 days), 19
months for Venus (some 23 days short of its actual 584 days), 8
months 20 days for Mars (only about one-third of its 780 days),
and 13 months for Jupiter and for Saturn (about midway between
their 399 days and 378 days respectively). The third sphere bore
the axis of the fourth sphere, offset by an amount differing from
planet to planet. The fourth sphere which bore the planet on its
equator rotated in the same period as the third but in the opposite
direction to it (see Fig. 4.4).

   Considering at first only the operation of the third and fourth
spheres, the planet, as seen from the centre where the Earth was
deemed to be, would trace a figure of eight, called a hippopede by
the Greeks. The long axis of the figure would lie along the ecliptic
on which points 1, 3, 5 and 9 in Fig. 4.5A would lie, whereas points

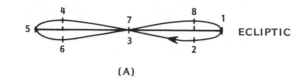

(A)

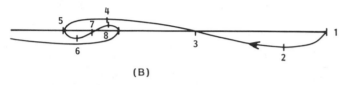

(B)

Fig. 4.5   (A) The hippopede or figure of eight generated by
the third and fourth spheres in Eudoxos's theories of the
star-like planets. (B) One part of the hippopede is drawn out
by the motion of the second sphere into a long phase of direct
motion and the other is compressed into a short period of
retrograde motion.

2 and 6 would be south of the ecliptic, and 4 and 5 would be north.
The effect of the rotation of the second sphere would be to draw
out the motion from 1, through 2, 3 and 4, to 5, and to compress
the motion from 5, through 6, 7 and 8, back to 1 as in Fig. 4.5B.
Having chosen an appropriate angle between the axes of the third
and fourth spheres, the planet as seen from the centre would have
a long segment of west to east (direct) motion, a short segment of
retrograde motion and some variation of latitude.

Appropriate displacements of the axes of the third and fourth
spheres can be found for Saturn and Jupiter in order to produce
fairly good retrograde arcs, although the implied variations in
latitude are in both cases too small. The observed data for Mars
cannot be accommodated. The greatest displacement, 90°, fails to
produce any retrograde motion but results in a variation of
latitude of 30°. A value can be chosen for Venus which gives a
reasonably good variation in latitude but not one that produces a
reasonable retrograde phase. A value can be chosen for Mercury
which gives a reasonable variation in latitude and a reasonable
retrograde phase (the arc of this phase is difficult to establish for
Mercury without sophisticated instruments). All in all, Eudoxos's
homocentric spheres constituted an ingenious model, enlightening

observations qualitatively but only sometimes quantitatively (see Aaboe 1974). It is to be doubted that Eudoxos took it to be other than a theoretical model. Had he regarded it as a physical reality he need not have repeated the outer sphere eight times. Further, there is no indication of how the model for one planet did not get in the way of that for another if they had physical reality. He did, however, seem to have regarded the spheres as crystalline, which suggests that if physical they were transparent and allowed the more remote fixed stars to shine through.

Aristotle treated Eudoxos's spheres as a physical reality where the motion of the outermost sphere was passed on through the differential 'gearing' of the outer mediate spheres in respect of period, direction and angle. Having got the properties of Saturn's motion as right as possible, they had to be brought back by appropriate 'gearing' to that of the outermost sphere so that Jupiter's four spheres could produce its observed properties of motion; this, Aristotle believed, could be achieved by reversing Saturn's 'gearing'. Thus Saturn's four spheres in the Eudoxan theory became eight in the Aristotelian, and so for Jupiter and for Mars; the Sun's three spheres had to be duplicated as had Venus's and Mercury's four each. Only the Moon's three spheres according to Eudoxos needed not to be duplicated. Thus, Eudoxos's 26 spheres were increased to 49 by Aristotle.

Kallippos, a generation after Eudoxos, improved on the Eudoxan theories by introducing some additional spheres. Thus for the Sun and for the Moon he introduced two additional contrary rotating but not markedly offset spheres. This enabled a reasonable account of the variation in the angular velocity of both bodies without sensibly affecting the Eudoxan variations in latitude (reasonable in the case of the Moon but insensible, and in fact unreal, in the case of the Sun). For Mars, Venus and Mercury, Kallippos introduced a further sphere between Eudoxos's third and fourth. This additional sphere rotated on an axis offset from that of the third in half the period but in the opposite direction. The inner sphere bearing the planet on its equator rotated on a further offset axis in the same period and the same direction as the third. This addition added to Eudoxos's hippopede a pair of loops at each end which should have been detectable during the retrograde phase of motion but were not observed. The sphere added by Kallippos gave both Mars and Venus a retrograde phase of reasonably good value, reasonably good variations in latitude, and in the case of Venus a good value for greatest elongation or angular separation from the Sun. Unfortunately it implied unobserved loops at the ends of the hippopede.

In both Eudoxos's and Kallippos's theories the retrograde phase for each planet would have been constant. It is in fact variable. Aristotheros in *c.* 300 BC drew attention to difficulties posed for the Eudoxan implication that the planetary bodies were at a constant distance from the Earth by (i) the varying brightnesses of Venus and Mars and (ii) the fact that solar eclipses are sometimes total, that is, the Sun is completely obscured, and sometimes annular, that is, the rim of the Sun remains visible.

We know through Ptolemy, who reported contentions of Hipparchos, that Timochares and Aristyllos, early in the third century BC, made measurements of the longitude of several fixed stars, observations which later enabled Hipparchos to establish the precession of the equinoxes and the approximate length of the tropical year. It is possible that Timochares and Aristyllos did not establish ecliptic longitudes but that Hipparchos converted their positional values in whatever terms they gave them to these terms.

It is probably worth noting that although various Greek observations were expressed above in degrees and minutes (360° in the circle, 60′ in the degree), this Babylonian convention from perhaps 500 BC may not have been adopted by Greek astronomers until the second century BC. For instance Eratosthenes found the obliquity of the ecliptic to be $\frac{11}{83}$ of a circle; one may justifiably wonder how the instrument he used, whatever it was, was graduated. Imagine the practical task of dividing a circle into 83 parts or even dividing say a quarter of it into 83 parts. The value, however, could be a compromise between a maximum observation of $\frac{48}{360}$ and a minimum of $\frac{47}{360}$ (but see Rawlins 1982); Archimedes closed in from above and below to find an approximate value of π. Eratosthenes may have done likewise.

HIPPARCHOS

Hipparchos, who flourished in
the middle years of the second century BC, was perhaps the greatest
ancient Greek astronomer. He was historically overshadowed by
Klaudios Ptolemaios, usually referred to as Ptolemy, second
century AD, who in addition to writing a comprehensive and
apparently definitive astronomical treatise plus supplementary
works, brought a number of Hipparchos's theoretical ideas to
adequate empirical application. We know what little we know of
Hipparchos's contributions mainly through citations and sometimes quotations in the writing of later authors, principally
Ptolemy. There are clear references to ten astronomical works
written by him; most, it would seem, were what we would call
papers rather than books. There are strong suggestions of a few
other astronomical works and in addition there are references to
or citations from works on mathematics (e.g. on chords in a circle),
on geography (where he criticized and tried to improve on
Eratosthenes's geography) and on astrology. Some 75 'fragments'
ranging from a sentence to a paragraph of Hipparchos's astronomical contributions have been preserved, as well as one small treatise.

Our best sources of reconstructed information on Hipparchos
are to be found in Dreyer (1905), Pannekoek (1961), Lloyd (1973),
Neugebauer (1975) and in *The Dictionary of Scientific Biography*
(Gillespie 1970–4). Hipparchos was born and grew up in the
Kingdom of Bithynia in northern Asia Minor. He lived later in
Nicaea where he seems to have begun his work, partly astronomi-
cal but probably mainly meteorological and astrological. Perhaps
before 141 BC he moved to Rhodes where the greater part of his

astronomical work seems to have been done. Three, probably early, works were 'on simultaneous risings' (known only from its title), 'on the rising of the twelve signs of the zodiac' (known from one fragmentary citation), and *The Commentary on the Phenomena of Aratos and Eudoxos*, which has been preserved in its entirety and which is the only full work of Hipparchos we have.

Though a more detailed examination of *The Commentary* will be needed later, a few initial remarks may be justified here. In the fourth century BC, Eudoxos had written an astronomical work *Phainomena*, now lost but which apparently gave positional information about a number of fixed stars in terms of simultaneous risings, settings and culminations (crossing the meridian passing through the poles and the zenith) and the like. Aratos, third century BC, incorporated much of this information in a poem which was widely read. Hipparchos in the second century BC set out to correct the errors in the positional information given by Aratos, presumably derived from Eudoxos. Hipparchos was a niggling critic; this is a fortunate feature because his criticisms of Aratos's data suggest that when he wrote *The Commentary* he had not yet discovered the precession of the equinoxes which he may have done by about 128 BC. Had he known about precession he might have regarded some Eudoxan positional values as being in apparent error as a result of that phenomenon. In the following two to five years, four other important papers or books from his hand appeared. One was on the length of the year, one on intercalary months and days, one on the system of the fixed stars (presumably his star catalogue, mentioned by Ptolemy and others but preserved if at all only in somewhat dubious or corrupt fragments unless Ptolemy's catalogue is based on it).

The second part of *The Commentary* is not devoted to correcting errors in the data of Aratos (and presumably those of Eudoxos) but to setting out a great deal of information about risings and settings of leading and trailing stars in the several constellations, simultaneous meridian crossing by several stars, angular extent of constellations and the like. Though the data in *The Commentary* are positional they never use the ecliptic co-ordinate system used by Ptolemy about 265 years later in his star catalogue, an issue which will by deferred for later discussion.

As we learned in the preceding chapter, Hipparchos improved on the Metonic cycle already improved in part by Kallippos.

Hipparchos also adjusted the eclipse cycle of 223 synodic months which the Babylonians had proposed probably before 500 BC. As we have seen, the recurrence of a pattern of eclipses depends on the lengths of the synodic month, the draconic month and the anomalistic month, all values well established by the late Babylonian astronomers, roughly contemporary with Hipparchos; 223 synodic months with a mean value of 29.53059 days total 6585.32 days, 242 draconic months with a mean value of 27.21222 days total 6585.36 days, and 239 anomalistic months of 27.55455 days total 6588.54 days. All three cycles are approximately the same but after several rounds they begin to fall out of step. Hipparchos suggested much longer cycles of 4267 synodic months (126 007.02 days), 4630.5 draconic months (126 006.18 days) and 4573 anomalistic months (126 006.95 days). He also suggested cycles of 5458 synodic months (161 177.96 days), 5923 draconic months (161 177.97 days), and perhaps 5849 anomalistic months (161 166.56 days).

Hipparchos seems to have demonstrated from Babylonian eclipse tables that whereas occasional lunar eclipses were separated by five months (presumably penumbral eclipses) as well as the more usual six months, some solar eclipses could be separated by five or seven months as well as the more usual six months.

Hipparchos is said to have established the apparent sizes of the Sun and the Moon but we do not know the values he obtained. We do have his values for lunar parallax established by two different procedures based on eclipse observations. From a solar eclipse which was total at the Hellespont but only four-fifths at Alexandria he worked out a value for the lunar parallax. From a lunar eclipse he worked out a parallax by measuring the size of the Earth shadow where the Moon passed through it. He concluded that the Moon's distance lay between 59 and 67$\frac{1}{3}$ times the radius of the Earth. These values are about 55 to 58 per cent of modern values, an underestimate which is not too defective if allowance is made for what were probably his rather crude instrumental aids. Hipparchos probably had instruments like those described by Ptolemy almost three hundred years later. These may have included graduated sextants and quadrants, the equatorial armilla, complex armillary spheres and the planispheric astrolabe later further developed by the Arabian astronomers.

Hipparchos improved on Kallippos's determination of the

intervals between equinoxes and solstices. As we have seen, Euktemon probably determined only the dates to the whole day of the solstices whereas Kallippos may have determined the dates of the equinoxes as well. Even so the latter was determining the length of the seasons only in whole days. Hipparchos found the intervals to be 94.5 days (spring equinox to summer solstice) 92.5 days (summer solstice to autumn equinox), values in excess by 0.49 days and 0.16 days respectively on modern calculations. Shortly afterwards Hipparchos found that the interval between, say, one spring equinox and the next was less than 365.25 days by about $\frac{1}{300}$ of a day, that is 365.2467, which is in excess by 0.0044 days (a little over six minutes). Thus while Hipparchos found the interval between the spring equinox and the following autumn equinox to be 187 days (0.65 days in excess) his findings implied an interval of 178.2467 between the autumn equinox and the following spring equinox (a shortfall of 0.6456 days).

Hipparchos had found values for the tropical year (interval of the return of the Sun to a given equinox or solstice) and for the sidereal year (interval of the return of the Sun to a given star near the ecliptic) by comparing his observations of the angular separations of stars (Ptolemy mentions Spica and Regulus) from equinoctial or soltitial points with similar observations made by Timochares and Aristyllos some 160 years before him. The stars reported had apparently shifted almost 2° eastward relative to the equinoctial and solstitial points. According to Ptolemy, Hipparchos at first thought that this might be an eastward shift of only zodiacal stars but later came to consider that it was a general phenomenon (see Fig. 5.1). The effect of this precession of the equinoxes, the westward motion of the equinoxes relative to the fixed stars, was to differentiate the tropical from the sidereal year. Some of Hipparchos's own observations seemed to vary from time to time (probably the result of errors of measurement). He may have been even less confident about the reliability of his predecessor's observations. Ptolemy reports that Hipparchos considered that precession occurring at a constant rate must have, on the evidence available to him, a lower limit of 1° per century. If we take Hipparchos's reported positional values more literally an upper limit would be 1° in about 78 years, a value Petersen (1966) argues Hipparchos set. The best modern value, apparently not subject to secular changes, is 1° in 71.71 years.

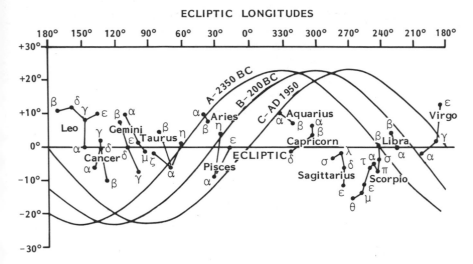

Fig. 5.1 *The westward shift of the equinoxes and solstices amongst the fixed stars (the precession of the equinoxes). The curves A, B and C represent the shifting celestial equator. In 2350 BC the spring equinox occurred in* Taurus, *in 250 BC it occurred slightly west of* Aries *and in* AD 1950 *it occurred in* Pisces. *Only the main stars in each zodiacal constellation are shown. The ecliptic longitudes shown at the top of the figure are those for* AD 1950, *for the earlier dates they would be shifted to the left by 30° and 60° respectively.*

Hipparchos concluded, as stated above from the evidence on precession, that the tropical year (from one spring equinox to the next) was $\frac{1}{300}$ of a day less than 365.25 days, namely that it was 365.2467 days, which was 0.0044 days (or 6.3 minutes) too long, and that the sidereal year (the interval between successive conjunctions of the Sun with a fixed star) was $\frac{1}{144}$ of a day greater than 365.25 days, namely that it was 365.2569 days, which is about 0.0005 days (or 0.78 minutes) too long. Distinguishing between the tropical and the sidereal years when there is so little difference between them was a great achievement. The close approximation to the actual periods is perhaps even more remarkable. Just as Hipparchos was uncertain about the constancy of the rate of precession, he was uncertain also about the constancy of the lengths of the tropical and sidereal years. Ptolemy asserted their constancy, again citing his own alleged observations. Actually the tropical year is very slowly decreasing and the sidereal year still more slowly increasing. These changes, however, are so slow that it would have been impossible for Ptolemy to have detected them from the short period of observation available to him, say 266 years from Hipparchos's observations to his own.

What has been reported so far about Hipparchos's astronomical contributions has involved observations and analyses of observations. He is also credited with the production of a star catalogue. From a comment of Ptolemy's this may have been produced in or more likely shortly after 127 BC, not long after Hipparchos's discovery of the precession of the equinoxes. The relation of the star catalogue in Ptolemy's *Syntaxis mathematike* to the lost star catalogue of Hipparchos is a matter of deep controversy which can scarcely be settled without better information than we have about Hipparchos's catalogue. One contention is that Ptolemy's catalogue is Hipparchos's catalogue incorrectly brought up to date in respect of longitude for precession in about 265 years by 2°40′ instead of 3°40′ (approximately) as a result of Ptolemy's underestimate of the rate of the precession of the equinoxes. There are, however, other possibilities. There may have been an error of about 1° in Ptolemy's setting of the armillary sphere with which he measured longitude or he may have made a mistake of 1° in establishing the spring equinox (the latter would have been a surprising error).

There are several arguments in support of the independence of Ptolemy's catalogue from the lost catalogue of Hipparchos, a difficult thesis to sustain in the absence of Hipparchos's catalogue. One such argument depends on the assumption that the Hipparchan catalogue would have given positional information not much different from that given in *The Commentary* written by Hipparchos perhaps only a few years earlier. *The Commentary* is primarily concerned with constellations and with the stars in the leading and trailing edges of these constellations in respect of risings, settings and meridian crossings, and especially with simultaneous risings, settings and meridian crossings, whereas the Ptolemaic star catalogue gives (in degrees and fractions of degrees) (i) ecliptic longitudes of the listed 1022 stars measured from the first point of the relevant sign of the zodiac and (ii) ecliptic latitudes. Neugebauer (1975) suggests that Hipparchos did not use ecliptic longitudes and latitudes. Yet Ptolemy in *Syntaxis mathe-*

matike cites two longitudes and one latitude established by Hipparchos and suggests that these are examples of several other Hipparchan values. Ptolemy also states that Hipparchos suggested that while precession affected stellar ecliptic longitudes (based on the spring equinox) it did not affect ecliptic latitudes, a hypothesis Ptolemy claims to have confirmed. Ptolemy cites Hipparchos as stating that observations by Timochares and Aristyllos were given in degrees of longitude. Degrees (360 in a circle) were probably not in use in the days of the two earlier astronomers. It is sometimes suggested that Hipparchos converted observations of his predecessors into degrees of longitude. It is possible but less likely that Ptolemy converted Hipparchos's values in whatever terms they were given into degrees of longitude.

Neugebauer (1975), in arguing that Ptolemy's catalogue was independent of Hipparchos's, attached much weight to evidence produced by Boll (1901), Vogt (1925) and Gundel (1936, cited by Neugebauer 1975). Boll's and Gundel's evidence is derived from medieval manuscripts alleging information which may be derived from Hipparchos's catalogue. Boll, for example, studied a list of 'Stars of Hipparchos' given in a fourteenth-century AD manuscript. It named 42 constellations and stated the number of stars in most of them. The numbers fall short by 17 per cent of the numbers in Ptolemy's catalogue, though the differences in missing numbers vary markedly from constellation to constellation, a variation that does not seem to be a function of the number of very bright stars in the constellation specified, assuming it to be like in content to the similarly named Ptolemaic constellation. Allowing for constellations omitted or constellations for which a star count is not given, Boll suggested that Hipparchos's catalogue could have contained no more than about 850 stars compared with Ptolemy's 1022. The accuracy of this account of Hipparchos's stars, recorded some sixteen centuries later, is dubious. It is more likely than not to be erroneous as a result of errors in successive reporting. A further point may be made about the material Boll examined. In Ptolemy's catalogue 99 stars are stated to be near or around the 'figures' (*imagines*) of the named constellations but not in them, leaving 913 in the constellations. Perhaps Hipparchos also had some stars outside the figures of the constellations. If so the estimate of 850 in the constellations would still fall short of the number in the catalogue but to a smaller extent.

The material studied by Gundel may also be subject to serious error (I have not been able to consult Gundel's paper and have had to rely on Neugebauer's summary). It was recorded centuries later than both Hipparchos and Ptolemy. It lists a number of stars and gives longitudes, most in whole numbers, which would have been on average approximately correct for Hipparchos's times. When 2°40' is added to these given longitudes Ptolemy's individual longitudes are not reproduced; this suggests that if the material examined by Gundel were derived from Hipparchos's catalogue, Ptolemy's catalogue is independent.

Vogt, using the information in Hipparchos's *Commentary*, was able to calculate ecliptic longitudes and latitudes for 122 stars. On average, the longitudes are about 2°40' less than Ptolemy's whereas the latitudes on average are about the same as Ptolemy's. There is, however, a wide disparity when the calculated Hipparchan values and the Ptolemaic values are compared star by star. The longitudes differ by from +2°54' to +6°36' before 2°40' is subtracted from the Ptolemaic values and the latitudes differ by from +2°54' to −2°32'. Clearly the Ptolemaic longitudes and latitudes of these 122 stars were not obtained by calculation from the positions given by Hipparchos in *The Commentary* and then erroneously corrected for precession. Further, it has been shown that when the positions of these 122 stars are calculated from modern considerations for the epochs of Hipparchos and of Ptolemy, the values in the Ptolemaic catalogue are appreciably more accurate than the values in Hipparchos's *Commentary*.

Vogt's argument seems to assume that Hipparchos in producing his catalogue would have used whatever positional information he had used in *The Commentary*. An interval of a few years seems to have elapsed between the writing of these two works but we cannot be sure how long it was. Almost certainly in this interval Hipparchos discovered the precession of the equinoxes. When he wrote *The Commentary* he was wanting to correct errors in the information from Eudoxos being broadcast by Aratos. The primitiveness of the positional concepts on which Neugebauer remarks may well be the result of Hipparchos's adhering to the terminology of Eudoxos. Having discovered the precession of the equinoxes he may have recognized that ecliptic longitude and latitude, with which he may have been familiar from the late Babylonian solar and lunar ephemerides, would be better in

Table 5.1 Frequencies of fractions recorded in Ptolemy's catalogue for stellar longitudes and latitudes

| | Number of instances | | |
| | Ptolemy | | |
Fractions	Longitudes	Latitudes	Theoretical distribution
0	226	236	171
$\frac{1}{6}$	182	106	128
$\frac{1}{4}$	4	88	86
$\frac{1}{3}$	179	112	128
$\frac{1}{2}$	88	198	171
$\frac{2}{3}$	246	129	128
$\frac{3}{4}$	0	50	86
$\frac{5}{6}$	102	107	128

The distribution of the fractions for the latitudes, though a poor fit to the theoretical distribution assuming a random distribution, is very much better than that for the longitudes; the value of chi-square for the former is 53.32 whereas for the latter it is 379.33 (the higher the value the poorer the fit).

charting stellar positions. He appears to have used terrestrial equatorial longitudes and latitudes in his geographical work and may well have seen the value of using celestial ecliptic longitudes and latitudes in charting stellar positions.

Ptolemy claimed that he (Ptolemy) included in his catalogue all stars down to the sixth magnitude visible to him. He says that he made his observations in Alexandria, yet he included no stars not visible from Rhodes where Hipparchos almost certainly compiled his catalogue. As Alexandria is about 5° south of Rhodes, Ptolemy would have had about 5° more of the southern sky to observe; this segment of the southern sky in his epoch included an appreciable number of fairly bright stars and several dozen down to the sixth magnitude, yet he includes none of them in his catalogue.

Several commentators (e.g. Dreyer 1917–18 and Newton 1977) have drawn attention to a puzzling feature in the frequency distribution of the fractions of a degree in Ptolemy's catalogue. He used sixths, quarters, thirds and halves both as unit fractions such as $\frac{1}{6}, \frac{1}{4}, \frac{1}{3}$ and $\frac{1}{2}$ and other fractions such as $\frac{2}{3}, \frac{3}{4}$ and $\frac{5}{6}$. The frequency distributions of these fractions in Ptolemy's catalgue are quite different for longitudes and for latitudes (see Table 5.1). The fractions used would result in an uneven random distribution of frequencies (see column 4 in Table 5.1). One might expect with copying and recopying an increase in values lacking a fractional value and a resultant decrease in neighbouring values ($\frac{1}{6}$ and $\frac{5}{6}$) and a similar effect around $\frac{1}{2}$. The basic pattern overlaid by these expected shifts in the frequencies of fractions can be seen in relation to Ptolemy's latitudes. Assuming an original distribution like that for latitudes, the addition of say 2°40′ to the longitudes would produce a distribution which resembles Ptolemy's frequencies for latitudes, whereas in fact it is markedly different.

All in all there seems to be a fair but far from conclusive case that Hipparchos's catalogue is preserved through an erroneous updating of longitudes allowing for precession in Ptolemy's catalogue. This controversial thesis could be confirmed or denied only if we had Hipparchos's catalogue or a reliable segment of it. It is not reasonable to assume that it was identical with the stellar positional information in *The Commentary* or that late medieval manuscripts which claim or seem to refer to some features of Hipparchos's catalogue are to be taken literally.

Finally Hipparchos is to be commended for his attempt to apply Apollonios's analysis of solar and lunar apparent motions into excentric circles or deferent-epicycles systems. Apollonios seems to have recognized that anomalous planetary motion (the speeding up and slowing down in the case of the Sun and the Moon, or the phases of direct, or west to east, motion interposed by shorter phases of retrograde, or east to west, motion in the case of the five star-like planets) could in principle be accounted for by means of excentric circles or by means of deferent-epicycle systems. Apollonios had demonstrated that the two models involving component uniform circular motions were equivalent in some circumstances as bases of analysis of anomalous (uneven) motion (see Fig. 5.2).

Hipparchos accounted effectively for the single anomaly of the Sun by placing the Sun in uniform angular velocity around a circle from the centre of which the Earth was offset. The amount by which the Earth was offset was in his account $\frac{1}{24}$ of the radius of

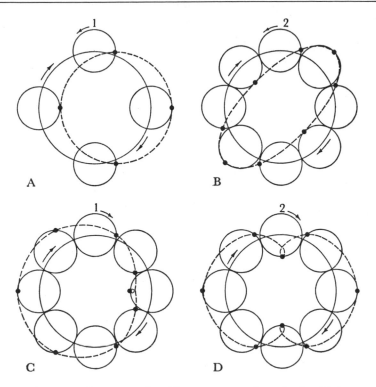

Fig. 5.2 *Figures produced by a point on the epicycle which is rotating in the same or the opposite direction to the deferent and at various relative rates. Note that in A the figure yielded is an eccentric circle.*

the circle and in the direction of perigee at 245°30′ in the order of the signs from the spring equinox (see Fig. 5.3). This accounts quite effectively for the solar anomaly including his assessment of the inequality of the seasons (94½ days for spring, 92½ days for summer, and probably 88⅛ days for autumn and 90⅛ days for winter).

For the Moon Hipparchos used a circle inclined by 5° to the ecliptic and shifting its inclination in the retrograde direction so that the lunar nodes completed a circle in 18⅔ years. That circle carried the centre of an epicycle; this centre moved directly in a sidereal month (see Fig. 5.4). The Moon borne on the circumference of the epicycle moved retrograde around it in the anomalistic

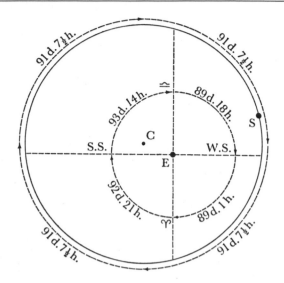

Fig. 5.3 *Hipparchos's explanation of the anomaly of the Sun. The Sun travels on the circle centred on C which is offset from the Earth, E, thus yielding unequal seasons.*

month. The ratio of the radius of the epicycle to that of the deferent was set at 0.0875.

These analyses were found to give good positional predictions for the Sun and for the Moon when the latter was at the syzygies (that is, when the Moon was in conjunction with or in opposition to the Sun). For some quadratures (Moon at first and third quarters, that is separated from the Sun by 90°) the predictions were equally good but sometimes they were in error by up to 2°39′. Hipparchos seems not to have been able to locate the circumstances distinguishing the good from the defective predictions at quadratures.

Though Hipparchos could see the general principle on which a deferent-epicycle system could be used to account for the anomalies of the star-like planets including both direct and retrograde apparent motion, he recognized that he lacked the quantitative observational data to work out the details.

All in all, his was a remarkable set of achievements; it is also remarkable that so much can be reconstructed when apart from one work we have only 'fragments' or brief references to his work by others from which to work.

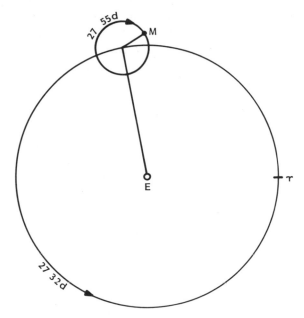

Fig. 5.4 *Hipparchos's attempt to explain the anomaly of the Moon by means of a deferent and an epicycle. The centre of the epicycle was held to rotate anticlockwise in a mean sidereal month and the Moon to rotate clockwise around the epicycle in a mean anomalistic month.*

The indirect information we have on Hipparchos leaves no doubt that he was a highly competent, technical, quantitative astronomer both in respect of observation and inference from observations and in respect of trying to formulate explanatory or at least predictive mathematical models. He may not, however, have been typical of his times. Some two centuries later Geminos wrote *Eisagoge eis ta phainomena* (or án *Introduction to the Phenomena*) which suggests what astronomical information it was thought desirable to expound to intelligent readers or perhaps students. This textbook is descriptive rather than explanatory and qualitative rather than quantitative. Now and again Geminos gave an explanation in general terms, for example a lunar eclipse occurs when the Moon falls in the Earth's shadow, or cited the value of some period such as the length of the month or of the year. He distinguished (i) synodic and anomalistic months, said to be 29.53

days and 27.5 days respectively, and (ii) solar and lunar years, said to be 365.25 and 354 days respectively; both pairs of values are poorer than those of Hipparchos. He also reported the Hipparchan differences in the length of the seasons.

Geminos distinguished between the twelve zodiacal constellations of various longitudinal lengths and the twelve zodiacal signs each of 30° in longitudinal extent. He seems not to have known that through precession the former were slipping eastward out of the latter. He identified some twenty-one constellations north of the zodiac and sixteen south of it. Most of these constellations have been preserved but often with changed figures. Some in the north have been changed and many have been added to those in the southern skies.

Geminos distinguished two kinds of physical astronomical spheres – one solid, apparently representative of the stars in the sky, that is, a celestial globe, the other apparently an armilla which had observational uses. He specified several celestial circles such as (a) the equator, the tropics and the arctic and antarctic circles, (b) the meridians, (c) the ecliptic and (d) the colures, the two great circles passing through the equinoctial points and the solstitial points respectively; the solstitial colure passes not only through the solstitial points but also through the ecliptic and equatorial poles. He also set out values for maximum sunlight to night at various terrestrial latitudes.

Geminos discussed several luni-solar near-equivalences such as the eight-year cycle, *octaeteris*, the nineteen-year (Metonic) cycle and the 74-year (Kallippic) cycle, and looked for the lowest common multiples in whole days of synodic months and anomalistic months. A luni-solar cycle is important for establishing the intercalation of a thirteenth month in some calendar years. Geminos also discussed in very general terms the patterns of risings, paths and settings of both the fixed stars and the five star-like planets (but the latter in the most sketchy terms).

Geminos also distinguished several climatic zones of the Earth and repeated earlier inferences about the diameter of the Earth. He denied the alleged effects of the stars on terrestrial weather and on human affairs. Interestingly he also denied that the fixed stars were equidistant and instead asserted that they were located at different distances but did not explain how this could be accommodated in a geocentric scheme.

The next preserved general treatise, Ptolemy's *Syntaxis mathe-matike*, is markedly different in most respects from Geminos's *Eisagoge*. Ptolemy's treatise is more technical, more quantitative and more detailed than Geminos's sketch. It clearly takes up the tradition pursued by Hipparchos some 250 years earlier and does not fall back on a vaguer, less precise quantitative analysis.

CHAPTER SIX

PTOLEMY

Klaudios Ptolemaios, or to give him the modernized form of his surname, Ptolemy, seems to have been a person of Greek descent working in or near Alexandria, Egypt, in the second century AD. All his preserved writings are in Greek, yet the Latinized form of his name, Claudius Ptolemaeus, suggests that he may have been a Roman citizen in what had become a Roman province almost two centuries before. Little is known of his personal biography, such as the dates of his birth and death, his familial antecedents, where and from whom he acquired his basic knowledge of mathematics, astronomy, astrology and geography. Quite often his acceptance as the authority in several of these fields led to the loss of statements of the contributions of his predecessors. Apart from Hipparchos's *The Commentary on the Phenomena of Eudoxos and Aratos*, our main knowledge of Hipparchos's astronomical contributions comes from what Ptolemy reports in his major astronomical treatise, *Syntaxis mathematike*. There are many references to pre-Ptolemaic astronomical writings in this treatise but few of the originals have been preserved and not much more of the detail. There are now two English translations of *Syntaxis mathematike*, one by R. C. Taliaferro (1935) and a second by G. J. Toomer (1984); the second is said by reviewers who are recognized authorities to be remarkably superior.

Three issues in relation to *Syntaxis mathematike* will be examined. First, the treatise set forward the geocentric argument that the Earth was an immobile body at the centre of the universe consisting of the other celestial bodies which revolved around it.

This so-called geocentric theory, as we shall see, was when worked out geofocal rather than geocentric. Second, the treatise proposed geometrical (mathematical) models with attached numerical values in order to account for the anomalous (uneven) motion in longitude of the Sun, the Moon and the five star-like planets. These mathematical models, trigonometrical rather than purely geometrical, though useful in making predictions of planetary positions did not necessarily provide a plausible account of physical reality, a point emphasized by some Arabian successors of Ptolemy and by some Renaissance astronomers in the Latin west. It will be necessary to examine in some detail Ptolemy's numerical geometrical planetary models. The Ptolemaic planetary models were improvements of Hipparchos's basically similar but often only vaguely sketched models. Though we have Ptolemy's own accounts of his planetary theories, his accounts are so technical and so archaic in style that only sophisticated readers can readily follow them. Therefore the general reader needs to begin with modern simplified expositions such as those of Dreyer (1905), Lloyd (1970), Neugebauer (1969) and Pannekoek (1961). Even Dreyer is hard sledding. Third, some attention will be given to the reliability of Ptolemy's reported observations which were so important in seeming to confirm his theoretical models. Finally, an account will be given of other contributions in other works by Ptolemy to areas related to astronomy.

———————— I ————————

There are several main constituents in Ptolemy's geocentric conception of the universe, including the fixed stars, the seven planets and the Earth. First, he claimed that the universe was a sphere, a view that seems to be supported by the apparent shape of the starry heavens. Second, he claimed the Earth to be a sphere, as is evident from its shadow as revealed in part during lunar eclipses. Third, he claimed that the apparently mobile celestial bodies revolved around the Earth, the fixed stars uniformly in a period of 24 hours and the seven planets in more complex variable motions requiring the formulation of the planetary theories which will be stated below. Fourth, he claimed, on grounds to be stated below, that the Earth was at the centre of the universe. Fifth, he claimed that while the fixed stars and the seven planets were in uniform and in anomalous (uneven) motion respectively, the Earth was immobile.

Ptolemy's grounds for proposing the fourth and fifth of these propositions call for some elaboration. I shall mention two grounds for the proposition that the Earth is at the centre of the universe. First, heavy objects fall to the Earth on straight lines directed to its centre. On Aristotelian principles, heavy objects seek the centre and light objects such as fire fly from it. In the absence of any evidence of gravitation to other points or objects, the centre of the Earth was apparently the centre of the universe. Second, Ptolemy drew certain inferences from the hypothetical displacements of the Earth from the centre of the universe. He considered the implications of a supposed location of the Earth on the polar axis but nearer to one pole than to the other and the implications of a supposed location of the Earth off the polar axis. He argued, invalidly, that if the Earth is on the polar axis, but nearer one pole than the other, the horizon would not cut the celestial sphere into equal halves, which it obviously does. Were the sphere of the fixed stars finite and on average equally populated, then on these assumptions the horizon would not cut the heavens into two equally populated halves, assuming that northern observers could see the southern skies.

Ptolemy considered the two Aristarchan and Seleukan propositions about the motion of the Earth: daily rotation around its polar axis and annual revolution around the Sun. He rejected the diurnal rotation of the Earth west to east on the ground that unless the air shares the Earth's rotation the winds will blow consistently from east to west and carry the clouds in the same direction. Even if the air shared the Earth's rotation, objects thrown up would return to the Earth west of their point of launching. These physical arguments overlooked inertia, a concept formulated by Newton but adumbrated vaguely by Buridan, as impetus, and more explicitly by Galileo who wrongly assumed celestial inertia to be circular and not rectilinear. Were the Earth to move from one point in space to another, we would encounter the wrongly alleged difficulties arising from the assumptions that the Earth is not at the centre. Ptolemy might have mentioned the implication of the Earth's alleged revolution around the Sun of the semi-annual parallax, which was recognized as a possibility in late medieval and early modern thought but not observed until the late 1830s.

A problem in all this is distinguishing between the motion of an observer and the motion of some external observed object. An observer on a moving boat will tend to see floating logs as travelling backwards relative to the (apparently) fixed position of the boat. If, as Ptolemy argued, the Earth is not in motion then it is apparent that the Sun, the Moon, the five star-like planets and the much more numerous fixed stars are in motion around the Earth. The fixed stars seem to be in uniform circular motion around the Earth. As these stars seem to be located on a celestial sphere, Ptolemy assumed that they were equidistant, their apparent brightness not being a function of their distances; we are now convinced that both absolute brightness and distance are relevant to apparent brightness.

Some modification of the uniform circular motion of the fixed stars had to be adopted for the variable apparent motions of the seven planets. Thus they had to be detached from the motion of the celestial sphere of the fixed stars and to be attributed to more complex explanatory models.

II

In relation to Ptolemy's planetary theories, the first thing to be emphasized is that Ptolemy had seven independent though analogous planetary theories, amongst some of which there were several coincidental features introduced in order to account for the joint apparent motions of the Sun, the Moon and the five star-like planets. He had in addition an eighth independent theory to account for the apparent motion of the fixed stars.

Ptolemy said that his observations confirmed Hipparchos's theory of the Sun. He was right in confirming Hipparchos's assumed degree of the excentricity of the Sun (though on present knowledge it is to be attributed to the excentricity of the Earth's orbit). He was wrong in claiming that in his time the direction of the apogee of the Sun (on a geocentric basis) was still as Hipparchos claimed. We know that it had in fact changed by 4°30′. We may reasonably doubt that Ptolemy made any relevant solar observations or, if he did, that he used the results to update Hipparchos's solar values. He did much better in respect of the Moon and the five star-like planets for which he generated new theories.

Ptolemy discovered the circumstance which resulted in the longitude of the Moon when at quadrature being sometimes

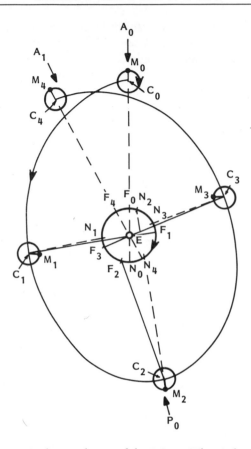

Fig. 6.1   *Ptolemy's theory of the Moon. The circle centred
on the Earth (E) and passing through F_0, F_1, F_2, etc., is the
'crank' which varies the distance of the Moon from the Earth.
The centres of the Moon's epicycles, C_0, C_1, C_2, etc., move
anticlockwise on the deferent which consists of a series of loops
because the apsides, the line joining the apogee, A_0, to the
perigee, P_0, rotates anticlockwise by about 26° in a synodic
month. The centres of the moving epicycle are equidistant from
the relevant points F_0, F_1, F_2, etc., on the 'crank' but are so
located that the lines N_0C_0, N_1C_1, etc., sweep over equal
angles in equal times. The Moon rotates anticlockwise around
the circumference of the moving epicycle. (The diagram is only
approximately to scale.)*

predicted correctly by the Hipparchan theory and sometimes predicted with an error of up to 2°39'. When the Moon was in syzygy (conjunction with or opposition to the Sun) the prediction from the Hipparchan theory was correct. At quadrature if the Moon was at apogee or perigee the prediction was also correct. But when the Moon was at quadrature and midway between apogee and perigee the prediction was in error by up to 2°39'. Ptolemy, in the course of trying to elucidate this anomalous situation, discovered that the lunar line of apsides (the line from apogee to perigee) was slowly rotating eastwards. He went on to generate an ingenious extension of Hipparchos's deferent-epicycle theory of the Moon. He put the centre of the Moon's deferent in retrograde motion around a small circle (a 'crank') centred on the Earth (E) (see Fig. 6.1). The radii C_0F_0, C_1F_1, C_2F_2, etc., are equal in length so the distance of C from E decreases and increases. The retrograde motion of C is such that it traverses equal angles in equal times not at E or at F but at N which is opposite to F on the 'crank'. The resultant deferent is not a closed circle but a looping oval.

This clever piece of mathematical analysis of apparent lunar motion enabled more accurate predictions of the longitude of the Moon at given times (the treatment here has ignored latitude). However, there were later objections to it by some of the Arabs and specifically by Copernicus. First, in referring equal angular velocity to the points N_1, N_2, N_3, etc., on the 'crank', which are not the centres of any circles, the analysis is in breach of the requirement for component uniform circular motion. Second, as it places the Moon at perigee at little more than half its distance at apogee, the apparent diameter of the Moon should be almost twice as great at perigee as at apogee, which it is obviously not.

The general pattern of the Ptolemaic theories of the five star-like planets was to have a deferent around which the centre of an epicycle moved in direct or anticlockwise motion in a sidereal year for Mercury and Venus and in the planet's zodiacal period for each of the other three. The planet was held to move in direct motion around the epicycle in the synodic period of the planet (see Fig. 6.2). The resultant motion of the planet is illustrated in Fig. 6.3 in respect of Mars.

Motion around the epicycle was reckoned by the ancients as angular departure of the radius of the epicycle to the planet from

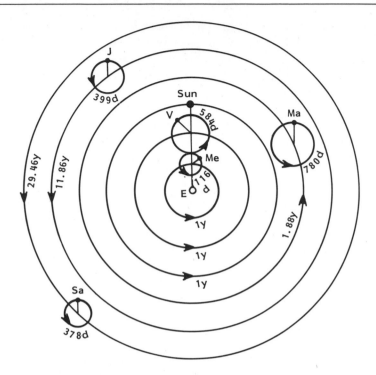

Fig. 6.2   *The major general features of Ptolemy's theories of the planets other than the Moon. Both the deferents and the epicycles move anticlockwise in the periods shown. The radii of the deferents of Mercury and Venus lie on the radius vector of the Sun and the radii of the epicycles of the three other star-like planets are parallel to the radius vector to the Sun. (Diagram not to scale.)*

an extension of the radius joining the centre of the deferent to the centre of the epicycle. Thus if the planet remained on the line joining the centres as in Fig. 6.4, it would be deemed not to have moved on the epicycle, whereas by modern reckoning it would be deemed to have made one revolution on the epicycle having moved through 360° from P_1 back to that point again. Reckoned in the modern way the period of the epicycle would be what we now know to be the heliocentric period of the planet in the cases of Mercury and Venus and a sidereal year in the cases of the other three planets. In this and in other respects such as the ratios of the

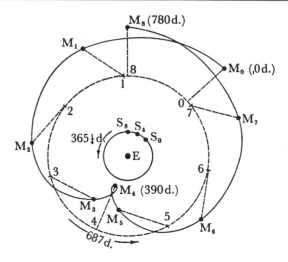

Fig. 6.3   The detailed features of Ptolemy's theory of Mars.
Beginning at time 0, the Sun and Mars are in conjunction;
390 days later at time 4, they are in opposition, the Sun having
made about $1\frac{1}{15}$ of a circuit and Mars a little over half a circuit
of the zodiac. At time 8, 780 days after time 0, they are in
conjunction again, the Sun having made about $2\frac{1}{7}$ circuits and
Mars about $1\frac{1}{7}$. In the neighbourhood of M_4, Mars is in
retrograde motion.

radii of the deferent and the epicycle, the deferents of Mercury
and Venus and the epicycles of the other three are equivalent to
the orbit of the Earth in a heliocentric treatment, and the epicycles
of Mercury and Venus and the deferents of the other three are
equivalent to the heliocentric orbit of the particular planet.
Ptolemy's ratios of the radii (epicycle to deferent) are shown in
Table 6.1, which also shows the ratios of the semi-major axes of
the orbit of Mercury and of Venus to that of the Earth and of the
orbit of the Earth to those of the other three.

The planetary theories each assumed that the Earth was offset
from the centre of the deferent. The motion of the centre of the
epicycle around the deferent was not uniform with respect to the
Earth or with respect to the centre of the deferent but, in the cases
of Venus, Mars, Jupiter and Saturn, was uniform with respect to
another point, labelled Eq in Fig. 6.5. This point in respect of which

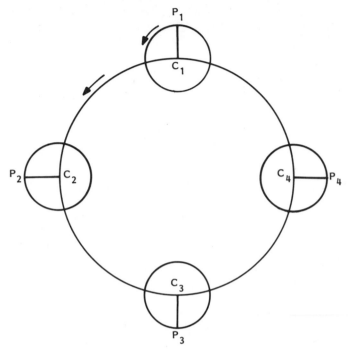

Fig. 6.4 *A deferent and an epicycle rotating in the same direction and in the same period in the modern reckoning. In the ancient mode of reckoning P would not have rotated as C rotated around the deferent.*

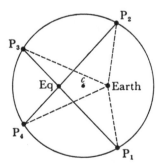

Fig. 6.5 *The relative positions of the centre of the deferent, of the Earth and of the equant. A point P has uniform angular velocity relative to the equant, Eq, whereas it has markedly variable velocity as seen from the Earth.*

Table 6.1   The ratio of the radius of the epicycle to that of the deferent in Ptolemy's theories compared with the ratios of the semi-major axes of the orbits of Mercury and of Venus to that of the Earth, and that of the Earth to those of the other three

	Ptolemy's ratios	Modern ratios
Mercury	0.3708	0.3871
Venus	0.7194	0.7233
Mars	0.6583	0.6563
Jupiter	0.1917	0.1922
Saturn	0.1083	0.1048

the motion of the centre of the epicycle was constant was called in later Latin accounts the *punctus aequans* or the equant. The ratios of the radii, the degree of offset from the centre of the deferent by the Earth, the location of the equant and the direction of apogee were chosen separately for each planet in order to yield, when taken with the period of the deferents and of the epicycles, which were based on observation, not assumption, the positions of the planets in longitude. The values of the radii of the deferents and of the epicycles were not observed whereas the values of the sidereal and synodic planetary periods were. The former were values selected so that when combined with the latter (observed) values the prediction of observed positions was enabled.

A more complex theory was developed for Mercury. The centre of the deferent was not fixed but moved around a circle in the opposite direction to but in the same period as the centre of the epicycle. The Earth was offset from the centre of this inner circle by twice the radius of that circle. The centre of the epicycle moved uniformly, with respect not to the moving centre of the deferent or to the centre of the inner circle or to the Earth but a point midway between the centre of the inner circle and the Earth (see Fig. 6.6). The analysis implied two perigees, contrary to presently accepted fact.

An unexplained coincidence in the Ptolemaic planetary theories is that the radius vectors from the Earth to the centres of the epicycles of Mercury and of Venus lie on the radius vector from the Earth to the Sun and the radius vectors from the centre of the

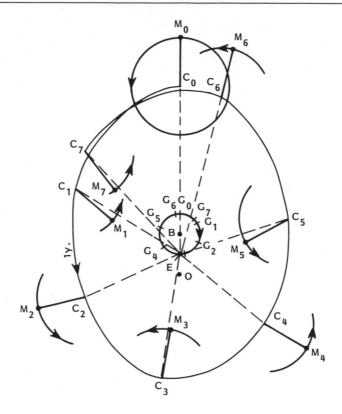

Fig. 6.6   *Ptolemy's theory of Mercury. B is the centre of a small circle (the 'crank'), E is the equant and O is the observer on Earth. C_0, C_1, C_2, C_3, etc., are in the centres of the epicycle and are located so that G_0O_0, G_1O_1, G_2O_2, G_3O_3, are equal in length and EC_0, EC_1, EC_2, EC_3, etc., traverse equal angles in equal times. The points G_0, G_1, G_2, G_3, etc., move around the 'crank' clockwise in one year, the points C_0, C_1, C_2, C_3, etc., move on the deferent anticlockwise in one year, and Mercury (M) moves around the epicycle anticlockwise in 116 days (the synodic period of the planet). At M_0, M_2, M_4 and M_6 Mercury is in what we call superior conjunction with the Sun and at M_1, M_3, M_5 and M_7 it is in inferior conjunction and in the middle of a retrograde phase. Only one epicycle is shown completely, the others are shown as arcs in order not to overcrowd the diagram.*

epicycle to the planet in the case of the other three are parallel to the Earth-Sun vector (see Fig. 6.2). With the assimilation of the deferents of Mercury and Venus and of the epicycles of the other three to the orbit of the Earth in a heliofocal theory, the coincidence disappears. It should be noted that in one respect Fig. 6.2 is misleading. It suggests that Ptolemy had a single planetary theory in which his several planetary theories formed a unified theory, whereas each Ptolemaic planetary theory is indeed completely independent of every other. Each stood or fell by itself alone.

In addition to accounting for planetary variations in longitude Ptolemy tried to approximate the variations of the planets in latitude relative to the ecliptic by giving the deferents and the epicycles tilts varying in amount from planet to planet. He had poor data on which to work and was much less successful in respect of motion in latitude than in respect of motion in longitude. With longitude he had greatest success with Mars, Jupiter and Saturn. Venus provided him with difficulties which he did not fully overcome, and Mercury with even greater difficulties. A problem for him in respect of variations in latitude was that he passed the plane of the deferent through the Earth whereas in the developed Keplerian heliofocal scheme the planes of the planetary orbits are passed through the Sun.

Though Ptolemy's arguments for the immobility of the Earth were physical, it is clear that with his deferents, epicycles and equants he was seeking a mathematical model for the purpose of calculating directions or positions of celestial bodies rather than a description of physical reality. An interesting case arises in his lunar theory. The radii he gives to the three component circles are such that the Moon's distance at apogee would be 1.9 times its distance at perigee. He claims in Book VI of *Syntaxis mathematike* that the Moon's diameter subtends at its greatest, presumably at perigee, 31'20". His variable distances imply at apogee a diameter of 16'50", which he would have known to be absurd. The accepted modern values are 33'36" at perigee and 29'30" at apogee with a mean of 31'7".

In respect of longitudes Ptolemy's tables based on his planetary theories enabled reasonably good predictions. With the passage of time the Arabian astronomers found them in need of correction; sometimes this resulted from errors in Ptolemy's basic data and sometimes from secular changes of which he could not always be expected to be aware. All in all his was a remarkable piece of mathematical modelling in pure terms and in generating useful tables.

——————————— III ———————————

In *Syntaxis mathematike* Ptolemy frequently cites positional, periodic and other values based on successive positions of the Sun, the Moon, the other planets and the stars. He cites quite a number of observations made by earlier astronomers such as Timochares and Aristyllos (both *fl.* 280 BC), Eratosthenes (276–194 BC) and Hipparchos (*fl.* 120 BC) in order to establish secular changes or to show the congruence of his findings with theirs. Ptolemy does not tell what sightings and angular measuring devices his precedessors used. He does, however, describe three instruments used by himself. One is a graduated quadrant which could by used for measuring angular differences up to 90° between two celestial bodies or between a celestial body and the horizon. A second is a ring set in the equatorial plane. The shadow cast by the upper part on the lower part of the ring enables the determination of the date of the equinoxes. The third device, which Ptolemy claims to have designed himself, he called an *astrolabon*, literally a star-taker. It was not a precursor of the later Arabian astrolabe but was what later acquired the name armillary sphere. It consisted of a pair of rings set fixed at right angles to one another to represent respectively the ecliptic and the meridian circle which passed through the two solstices and both the ecliptic and equatorial poles (this meridian is called the solstitial colure). This pair of rings may be turned in a fixed outer ring on two pins to the colure ring at the ecliptic poles. Inside the double rings a graduated ring may be rotated on two pins; from it may be read an ecliptic longitude. Sliding inside that ring is another graduated ring with sights which when trained on a star to enable the reading of latitude on that ring as well as longitude on the ring in which it slides (see Chapter 10). It would be surprising if Hipparchos some 260 years earlier did not have somewhat similar devices even in less developed form. Timochares, Aristyllos and Hipparchos, according to Ptolemy, measured the declinations of some eighteen stars, establishing values in degrees and fractions of a degree down to $\frac{1}{6}$, that is 10′

of arc or some equivalent fractions of a circle. To obtain such precise values they must have had some aid like Ptolemy's *astrolabon*.

For quite a long time doubts have been cast on the genuineness of some of the data Ptolemy claims to have established by his own observations. Tycho Brahe in the late sixteenth century doubted that the longitudes in Ptolemy's star catalogue, out on average by about 1°, were based on observations made by Ptolemy rather than on corrections made by formula to some earlier catalogue. Lalande in the seventeenth century questioned many of Ptolemy's alleged observations. Delambre, early nineteenth century, raised similar doubts on the catalogue and on some related matters. Dreyer (1905) and Pannekoek (1961) have extended the list of alleged observations that have constant errors not readily attributable to the setting of the instruments used by Ptolemy. Newton (1977, 1980) has produced a detailed case in support of the charge that Ptolemy fudged his own 'observational' data to fit his theories or his preconceptions. Other scholars (cf. Gingerich 1980, 1981) consider that Newton has been overzealous in his prosecution of Ptolemy. There will be little point in going into all the cases where Ptolemy's claims are suspect. It should suffice to take some examples of various kinds.

Eratosthenes, *c.* 225 BC, found the differences in solar altitudes at the summer and the winter solstices to be $\frac{11}{83}$ of a circle, that is about 47°42′39″, which gives a value for the obliquity of the ecliptic of about 23°51′19″.5. In Eratosthenes's epoch the correct value, on modern considerations, would have been 23°43′20″, so it may be said that Eratosthenes's value, less that 8′ in error, was a very good early estimate. The obliquity of the ecliptic has been slowly decreasing by about 0″.47 per annum, so that by Ptolemy's time the value should have been about 23°40′24″. Ptolemy, however, says that his observations confirm Eratosthenes's exactly! Yet the error of less than 8′ in Eratosthenes's value had increased to almost 10′57″ in Ptolemy's epoch, which should have been detectable by Ptolemy.

Ptolemy reported that Hipparchos, having compared the ecliptic longitudes of some stars established by himself with those established some 160 years earlier by Timochares and Aristyllos, found a shift of the equinoxes (to the west) of almost 2°. Hipparchos, according to Ptolemy, concluded that the precession of the equinoxes so revealed occurred at the rate of at least 1° in a century. Granted that the change was less then 2° in 160 years, as reported, the rate would have been in the neighbourhood of 1° in 78 years, which it seems Hipparchos may have set as an upper limit (see Petersen 1966). Ptolemy said that his observations of the ecliptic longitudes made some 266 years after Hipparchos's observations confirmed the 1° precession in a century suggested by Hipparchos as a probable value. He cited the longitudes of eighteen stars as measured by Timochares or Aristyllos, by Hipparchos and by himself. He selected, according to Newton, apparently at random, three stars with longitudes between the winter and summer solstices and three with longitudes between the summer and winter solstices. The differences between Ptolemy's and Hipparchos's longitudes for these six stars show a 1° per century precession. Had Ptolemy calculated the rate from the differences in longitude of the other twelve stars it would have been, claims Newton, 1° in about 70 years. Newton claims that in this case Ptolemy chose his alleged observations to fit his preconceptions.

These are not isolated examples of possible fudging by Ptolemy of alleged observational values. A few other instances will be cited from Newton's more extensive collection.

Ptolemy claimed that his observations confirmed Hipparchos's value for the excentricity of the Sun and Hipparchos's value for the direction of the solar apogee, both expressed in geofocal terms. While, on present knowledge, the excentricity would not have changed, at least appreciably, in the intervening 266 years, the direction of the solar apogee would have changed by about 4°30', which Ptolemy could have detected had he made reasonably accurate observations of the dates of the equinoxes and the solstices and adjusted Hipparchos's solar theory accordingly. In working out his solar tables, Ptolemy used Hipparchos's solar apogee of 65°30' instead of the 70° which obtained in his own time. This led to an error in his solar table and, as a result, in his other planetary tables, because he related the latter to his solar table; the error is about 4°30'. These erroneous values fitted Ptolemy's planetary theories more closely than would the values we now calculate for his epoch.

In order to check such periods as the tropical year Ptolemy cites the dates of several solstices and equinoxes said to have been observed by some of his predecessors and by himself. His own alleged observations are according to Newton about 28 to 36 hours too late as calculated from present considerations. This delay corresponds closely to Hipparchos's estimate of the tropical year ($365.25 - \frac{1}{300}$ days) which was slightly too long – about 6.3 minutes per year, or about 26 hours 24 minutes in 250 years, or about 31 hours 41 minutes in 300 years.

Again, in order to check on various solar-lunar periodic relationships, Ptolemy compared the dates and times of eclipses said to have been observed by himself in AD 133, 134 and 136 with the dates and times reported by several earlier astronomers. Ptolemy's 'observed' values differ by no more that an hour or two from the values which can be derived from his solar and lunar tables. From modern considerations, as expressed for example in Oppolzer's (1887) and Tuckerman's (1962, 1964) tables, Ptolemy's reported dates and times are over half a day in error.

It would be sad to think that so able a mathematical astronomer as Ptolemy dominated astronomy for 1500 years partly through his mathematical skill and perhaps partly through his fudging of alleged observational data selected to fit his mathematical models. His apparent finding of numerous observational data puzzled and confused many of his successors including the astute Arab astronomers and Copernicus.

As Hartner (1977) has pointed out, it was not until Tycho Brahe that we had an astronomer fully appreciative of the need for accurate observational data obtained through repeated observations made with instruments of high reliability. It is not necessarily heinous for an earlier astronomer to have been satisfied with a rather unreliable value or for him to round it to a value which was closer to some theory he held; perhaps it was not right for him to claim that he had observed something when he was merely agreeing with the observation of a predecessor. Ptolemy may not have been the villain Newton makes him out to be but he did confuse many of his successors.

IV

After writing *Syntaxis mathematike*, Ptolemy wrote (i) a work on astrology, *Tetrabiblos*, which will be discussed in a later chapter, (ii) a *Geographical Directory* giving the terrestrial locations (longitudes from the Fortunate Isles in the west, and latitudes north of the terrestrial equator) of numerous places and the latitudinal zones or

Table 6.2 Ptolemy's estimates of the minimum and maximum
planetary distances in terms of Earth radii

Moon	min.	33	Mars	min.	1 260
	max.	64		max.	8 820
Mercury	min.	64	Jupiter	min.	8 820
	max.	166		max.	14 187
Venus	min.	166	Saturn	min.	14 187
	max.	1 079		max.	19 865
Sun	min.	1 160			
	max.	1 260			

climata, and (iii) a set of *Handy Tables* giving more detailed celestial
positional information over a longer period of time than was
contained in the several planetary tables in the *Syntaxis*. In a later
work surviving only in part in Greek but more fully in an Arabic
translation (it is commonly known as *The Planetary Hypotheses*)
Ptolemy extended a concept introduced in the *Syntaxis*. Following
Hipparchos he thought he could calculate the distance (maximum
and minimum) of the Moon and the Sun in terms of Earth radii.
His estimates were too low, especially in the case of the Sun. In the
Syntaxis he indicated that if Mercury and Venus lay between the
Moon and the Sun, then the ratios of the radii of the deferent and
of the epicycle of each of those two planets which he needed to
assume, could be given a value in Earth radii, so that Mercury's
least distance was greater than the Moon's maximum distance,
Venus's least distance was just greater than Mercury's maximum
distance and Venus's maximum distance just below the Sun's least
distance (at perigee). In *The Planetary Hypotheses* Ptolemy extended
this nesting of spheres to Mars, Jupiter and Saturn. The minimum
and maximum selected by Ptolemy are set out in Table 6.2.

He concluded that unless there was empty space beyond
Saturn's maximum distance, the sphere of the fixed stars must be
at least 19 865 Earth radii distant. The nearest fixed star is in fact
3 200 000 times more distant than the maximum distance of
Saturn.

Even when medieval Europe had lost touch with the *Syntaxis*,
which it probably could not have understood, the *Tetrabiblos*, the
Handy Tables (used primarily for astrological purposes) and the
nesting of the planetary spheres continued to exercise considerable
influence.

THE AFTERMATH OF HELLENISTIC ASTRONOMY

After Ptolemy, Hellenistic mathematical astronomy went into a steady decline as did most other branches of Hellenistic science after the second century AD. In the Greek-speaking eastern part of the empire the state of affairs was rather better than in the Latin-speaking west. In addition to preserving Ptolemy's writings, as well as those of Appollonius, Archimedes, Aristotle and Euclid, commentaries on these texts were being written. For example, Pappos, early fourth century, who was primarily a mathematician, wrote a commentary on Ptolemy's *Syntaxis mathematike* concentrating on the explication of the mathematical methods of that work but adding little or nothing to the astronomical data and theories (see Rome 1931). He reported a solar eclipse in a way that suggests that he observed it but it is just a stray empirical fact. Theon (see Rome 1931), late fourth century, was more concerned with explicating the explanatory analysis in terms of excentric deferents, epicycles and equants. His long-winded verbal accounts suggest that by his day students in Alexandria, where he taught, had difficulty in comprehending Ptolemy's formal mathematical methods. He also wrote a long and a short commentary on Ptolemy's *Handy Tables*, demonstrating how to use them but not adding anything. He reported two eclipses which he says he observed from Alexandria – almost certainly the solar eclipse of 16 June AD 364 and the lunar eclipse of 26 November AD 364. In discussing the Hipparchan-Ptolemaic doctrine of the precession of the equinoxes, a reported phenomenon he seems to doubt, Theon mentions that certain ancient 'astrologers' held that the equinoxes advance by 8° and then retrogress by 8°. This theory of trepidation of which we have no other Greek account turns up in Arabian astronomy (often differing in detail).

Proklos, fifth century, though predominantly a philosopher, wrote an *Outline of the Hypotheses of the Astronomer*, in which he gives a summary version of Ptolemy's planetary theories. He recognizes the value of Ptolemy's objective in reducing anomalous planetary motion to component uniform circular motions and he admits the value of Ptolemy's analyses and parameters in calculating planetary positions. He says, however, that the analyses are so complex that he doubts that they can be an accurate account of what in fact happens, a doubt later expressed by several Arab astronomers. Proklos does not, however, suggest a more plausible realistic account (see *The Dictionary of Scientific Biography* on Proklos).

In the Latin-speaking West the state of affairs was distinctly worse. From the first to the fifth centuries AD accounts of astronomy were included in several encyclopaedias. In the first century Pliny the Elder had a long section on astronomy in his multi-volume work *Natural History*, in the third or fourth century Selinus had such a section in his *Collection of Memorable Facts* and in the fourth or fifth century Martianus Capella had an incomplete section in his *Marriage of Philology and Mercury*, an account of the seven liberal arts. Such works seem to have been based largely on a now lost compilation by Varro, first century BC (see Stahl 1945). These accounts of astronomy are predominantly descriptive and rarely explanatory. Their origin is pre-Ptolemaic though post-Hipparchan. The Hipparchan theory of the Sun's orbit being excentric to the Earth is used to explain the varying lengths of the seasons (or the anomaly of apparent solar velocity) and eclipses are explained with varying degrees of precision. The complex pattern of apparent motion of the star-like planets is described in correct qualitative terms but often incorrect quantitative terms; no hint of Ptolemy's deferents and epicycles is given. Martianus Capella, a fifth-century encyclopaedist, interestingly resurrected Heraklides's theory that Mercury and Venus revolve around the Sun while the latter revolves around the Earth. In general these commentators follow Eudoxos's designations of the constellations as reported by Aratos, though sometimes neighbouring constellations are amalgamated. They give information about constellation risings and settings simultaneous with those of the signs of the zodiac, and the lengths of daylight and night, season by season at several terrestrial latitudes. In citing periods of the year and of the sidereal and synodic months and the other planetary periods they tend to round the more precise Greek values; thus Hipparchos's tropical year of $365\frac{1}{4}-\frac{1}{300}$ days becomes 'a little less than $365\frac{1}{4}$ days' or even 'about 365 days'.

The early Christian Fathers contributed to the decline of Hellenistic astronomy and other branches of science in both the west and the east of the empire. After Constantine had adopted Christianity as the official religion (early fourth century) not only was paganism discouraged, if not suppressed, as a religion but also pagans had obstacles placed in their way in teaching or otherwise promulgating views on secular matters where these views seemed to be in conflict with the Scriptures. Extreme examples of the rejection of Hellenistic astronomy were provided by Tertullian (early third century), by Lactanius (early fourth century), and by Kosmas (sixth century). Without differentiating amongst the details of their several views it may be said that they rejected the Hellenistic notion of the sphericity of the Earth and of the universe in favour of a layered, flat, square scheme as suggested in Genesis. Indeed to varying degrees they tended to support the view that the Mosaic Tabernacle represented the shape of the universe. Rather than conceding that the Sun between sunset and sunrise passed underneath a spherical earth, such thinkers argued that at sunset it fell behind a mountainous wall and after passing south behind the wall, rose again in the east. They could not admit that there was a 'beneath' to their supposedly flat earth, just as they could not admit an antipodes inhabited by people who stood 'upside down'. Not all of the early and later Christian scholars should be saddled with the views just stated in what is probably a caricatured, because abbreviated, fashion. Several of them tried to reconcile scriptural and scientific cosmological contentions. Early examples were Philiponos of Alexandria (late sixth century) and Isidorus of Seville (early seventh century). Both reported the sphericity of the heavens and Isodorus added the daily rotation. Isidorus also claimed that the Moon was smaller than the Sun and much nearer to the Earth. He placed Mercury and Venus between the Moon and the Sun and the other planets beyond the latter. He gave periods of revolution for all seven planets but in general these were seriously wrong and could not have been immediately based on observations. Scarcely any of this small group of scholars who remembered something of Hellenistic astronomical conceptions could be said to have made an impression on what the Scriptures said but they were ready to look for compromise, which went further and further as the medieval centuries passed. There was a limit, however, to the compromises, as Galileo was to discover in the seventeenth century. His predecessors such as Bruno had not trodden warily enough.

The only matter where astronomy was highly relevant to practical ecclesiastical concerns was computing the date of Easter, and the practices adopted reveal the decline of astronomical knowledge in both the western and eastern empire, represented by Rome on the one hand and by Byzantium (earlier Antioch) and Alexandria on the other. There were two problems in determining the date of Easter. First, what was the relation of Easter to the Jewish Passover? The first day of Passover was the fourteenth day of the month Nisan, a month which began with the first visible crescent moon usually after the spring equinox. The fourteenth day of a month counted from the first visible crescent would be about the day of the full moon, so 14 Nisan would ordinarily be the day of the full moon after the spring equinox. Some early Christians favoured the first day of Passover as the date of what was later called Easter, because it was on this day (beginning at sunset) that Jesus held his last supper, at which he instituted the Eucharist, and was later tried and crucified. A more popular view was that emphasis should be placed on Christ's resurrection, which occurred on the first day of the Jewish week (or the pagan Sunday) after that Passover. Further, it came to be considered that the Christian festival should not coincide with a Jewish festival, so if Passover began on a Sunday, Easter was decreed to occur on the following Sunday. Second, in order to apply the later rule that Easter falls on the first Sunday after the day of the full Moon on or after the spring equinox, one needs to be able to specify or to predict the day in which the spring equinox occurs and to predict the day of the full Moon on or after that date. By the second or third century AD the Christian authorities assumed that the spring equinox occurred on 25 March (Julian). This was a seriously mistaken view. From the inauguration of the Julian calendar, in 45 BC, the spring equinox never occurred later than 23 March (this may be calculated from Tuckerman's tables) and because the average Julian year of 365.25 days is slightly longer than the tropical year, the date of the spring equinox advanced through the calendar at the rate of about 0.78 days per century. At the Council of Nicaea, AD 325, the assembled Fathers declared 21 March to be the date of the spring equinox. Actually in that year the date of the spring equinox was 20 March as it was in the two neighbouring years.

The predicted date of the full Moon was calculated by means of some cycle such as the Metonic cycle; nineteen years with seven years having thirteen months approximates 235 months. The discrepancy between nineteen such years and 235 synodic months is quite small but in a century accumulates to almost half a day. Hipparchos could predict spring equinoxes to within a few hours and his Babylonian contemporaries could predict full Moons to within about two hours. Through the greater part of the first millennium AD the Christian scholars were doing very much worse. The computed date of the spring equinox could be in error by a day or two, if not more, and the computed date of the full Moon could be in error by comparable amounts. Different church groups using different rules of thumb often celebrated Easter on dates a week or more apart. Attempts to remove this 'scandal', for example the Council of Nicaea in AD 325 and the Synod of Whitby in AD 664, merely preferred one arbitrary answer to another.

In the early centuries of the Christian era, the Middle Eastern churches, centred especially in Syria with Antioch as a major headquarters and in Egypt with Alexandria, began to diverge from the doctrines of the churches located in Rome and in Constantinople (Byzantium). These centres in Syria and Egypt became strongholds of Nestorian, Monophysite and other ultimately proscribed sects. Members of these sects, especially the Nestorians, began translating Greek religious texts into Syriac, a western Semitic language with its own alphabet. They followed by producing Syriac translations of Greek philosophical, mathematical and scientific treatises. Some of the works of Plato, Aristotle, Euclid, Apollonios and Ptolemy were translated into Syriac, an important fact when translations into Arabic were begun. It was easier to translate from Syriac to Arabic than from Greek to Arabic, because Syriac and Arabic are both Semitic languages, whereas Greek, like Latin, old Persian, Sanskrit and modern European languages, belongs to the different class of Indo-European languages.

Many Nestorians and some Monophysites under harassment from the Orthodox Church (the Roman-Byzantium schism had not yet occurred) tended to migrate to the Persian empire in Mesopotamia (modern eastern Syria and Iraq) and Iran, and maintained active schools of intellectual and scholarly discussion, including translation and commentary on Greek texts in Syriac (see O'Leary 1949).

Fortunately post-Hipparchan but pre-Ptolemaic Hellenistic astronomy had drifted across to India at about the beginning of the Christian era and flourished for the next few centuries (see O'Leary 1949; Pingree 1963, 1974b). A series of astronomical texts called *siddhanta* incorporating Indian astronomical knowledge were published from about AD 400 to about AD 1200. A few elements were indigenous and drawn from the long-standing traditions reported in the first millennium BC but most of it had a Hellenistic-Babylonian origin. Possibly indigenous but more probably an adaptation of a Babylonian construction was the set of 27, later 28, *nakshatra* stars, 'the brides of the Moon'. The Moon moves eastward amongst the fixed stars on average a little over 13° per day (because of the Moon's variable velocity the movement may be as little as about 11° or as much as about 15°). The Moon deviates in latitude as we have seen by up to about ± 5° from the ecliptic. The Indian *nakshatra* stars, ten of which were Babylonian reference stars, tend to deviate more from the ecliptic than did the Babylonian reference stars − about half of them deviate by more than ± 10° the limits of all the Babylonian reference stars; four were over 25° north or south of the ecliptic. The average longitudinal spacing of the *nakshatra* stars is a little over 13° but five spaces are less than 6° or greater than 20°. There is not always a sufficiently bright star near enough to the ecliptic or near enough to 13° from its predecessor and its successor. As the Moon circles the heavens in 27.32 days, the Moon would on a given evening be about 4° to the west of the 'bride' of 27 days before and on the 28th evening almost 9° to the east of her. It is obviously a very crude piece of astronomical referencing. Yet it bestows the month names in the Indian calendar: the lunar months, the solar months and the civil months. Thus the lunar month (*chandra masa*) *Kartikka* is so called because it would ordinarily have begun when the Moon was near the *nakshatra Krittika*, the Pleiades (see de Saussure 1919–20); one of these related pair of words is an adjective and the other a noun.

Interestingly, just as twelve segments of the ecliptic, each 30° in longitude, replaced in Babylonian astronomy the twelve zodiacal constellations of varying extents, so 27 *nakshatra* segments, each 13°20′ in longitude, ultimately replaced the unevenly spaced *nakshatra* stars.

The early *siddhanta* texts adopt the seven-day week with planetary day names; these are sometimes clearly corruptions of the Greek names although in Sanskrit there are long-established planetary names. The twelve signs of the zodiac were also adopted usually in northern India with Sanskrit names which are the equivalent of the Greek and thus the Babylonian names. There are a few changes: the Greek *Hydrochoos* (Latin *Aquarius*) became *Kumba*, 'the pot', the Greek *Aigokeros* (Latin *Capricornus*) became *Makara*, 'the sea monster' (it was 'the fish-goat' in Babylonia), and the Greek *Toxotes* (Latin *Sagittarius*) became *Phanus*, 'the bow'. In southern India, where Dravidian languages (not in the Indo-European language family) were used, the names adopted for the signs are clearly corruptions of the Greek: *Kriya* for *Krios* (*Aries*), *Tavuri* for *Tauros*, *Jituma* for *Didymos* (*Gemini*) and so on (see Chapter 12).

There are some values in first-millennium AD Indian astronomy shared with both Babylonian and Hellenistic astronomy, such as the Metonic cycle of 235 synodic months in nineteen years, the so-called Saros eclipse cycle spanning 223 synodic months with possible eclipses six months apart for series of seven or eight eclipses but only five months between successive series. Several Babylonian periods which were adopted by the Greeks, such as a synodic month of slightly more than 29.53058 days, are to be found. Hipparchos's theories of the Sun and the Moon were adopted as were deferent-epicycle theories for the five star-like planets, but these latter theories like that for the Moon are simpler than Ptolemy's, for instance the equant is not used. The length of the tropical year and the precession of the equinoxes are reported usually in Hipparchan form. First-millennium Indian astronomers tended to think of precession as coming to an end or increasing and then decreasing over a period of time (the latter is the Greek notion of trepidation which some Greeks, according to Proklos, supported and which some Arabian astronomers later formulated). Corrections to stellar longitudes as a result of precession were deemed by several astronomers to be unnecessary after dates, about which they disagreed, in the middle part of the first millennium AD.

There are some features of this medieval Indian astronomy that are Babylonian and not found in Hellenistic astronomy as it has been preserved for us. First, in some preserved documents there is clear use of Babylonian mathematical methods such as those using step functions or zigzag functions to calculate the extremes of variable planetary velocities. The Greeks often made use of values arrived at by the Babylonians using these methods but they did not seem to have used the methods themselves. Second, because it was difficult until late in the piece for the Babylonians to establish whether a future calendar month would have 30 or 29 days in it, it was difficult for them to specify the day of the month on which some predicted celestial phenomenon would occur. Ingeniously, they resorted to the concept of thirtieths of a month, one solar day in a 30-day month and 0.966 of a solar day in a 29-day month. We do not know whether the Babylonians had a name for these thirtieths of a calendar month. Modern scholars speak of them as 'lunar days' whereas the Indians after they adopted this Babylonian convention called them *tithis*, as they still do today.

The Indian astronomers added a few things to the melange of Babylonian and Hellenistic astronomy (and also perhaps to astrology) which they imported. First, in some of the planetary theories they assumed a varying radius for the epicycle which resulted in an elliptical epicycle. Nothing remains to us from Hellenistic theories suggesting a Greek origin for this. Ptolemy's theory of the Moon implies very marked difference in the distances of the Moon but he achieves this by means of his 'crank' for the centre of the deferent and not by departing from circularity of the lunar epicycle. Second, the Indians adopted a solar as well as a lunar month. The Indian lunar month, like the

Babylonian and the derived Jewish, Greek and Islamic lunar months, is based on the successive first visible crescent moons after sunset (at first as observed but later as calculated). The Indians added a set of twelve solar months marked by the entry of the Sun into successive signs of the zodiac. The Indians called these *saura masa* (solar months); the shortest consisted of 29 days, most were 30 or 31 days and one was 32 days in length, reflecting the variable apparent velocity of the Sun.

It is not easy to specify the sources of this mixture of Babylonian and Hellenistic astronomy to be found in India from about AD 400 onwards. It may have come through the Greek Bactrian kingdom which survived in the northwest long after the Parthians had subjugated Hellenized Babylonia. It may have come from Hellenized Egypt (prior to Ptolemy the astronomer) but in that case its transfer was incredibly slow. It could have come from Babylonia itself, still Hellenized despite the displacement of the Seleucid rulers. Whatever the source (see O'Leary 1949), late in the first millennium AD a great deal of pre-Ptolemaic Hellenistic astronomy had been preserved in India and modified, except in just a few respects, in not very substantial ways.

In AD 773 there appeared in the court of Caliph al-Mansur an Indian scholar who could, using a *siddhanta* he had brought with him, calculate by means of mathematical tables contained in it the dates of possible eclipses and other celestial phenomena. Al-Mansur ordered that the *siddhanta*, presumably in Sanskrit, be translated into Arabic.

The strong hints of a Greek origin of much Indian astronomy such as embodied in the *siddhanta*, led the Arabian scholars to look for Greek texts on astronomy and mathematics, and later other scientific works on physics, chemistry, biology and medicine, which they translated into Arabic or less frequently into Persian. Some of the available texts had been translated, as already reported, from the Greek into Syriac. As Greek science was declining, many earlier texts were translated into Syriac, almost as a result of a sense of duty by some dissident Syriac-speaking Christian groups such as the Nestorians and the Monophysites who had suffered under Byzantine orthodoxy.

From the ninth century works by Ptolemy, Aristotle, Apollonios, Archimedes, Euclid and others had been translated into Arabic. In most cases Greek texts have been recovered today in the West but in a few cases we are dependent on the Arabic translations from Greek or from Syriac for Greek texts which have not survived. Further, we have Arabic commentaries on the Greek treatises. Later a more detailed account of so-called Arabian astronomy will be given.

In the thirteenth, fourteenth and sixteenth centuries scholars in the West who used Latin as their medium began translating into Latin Greek texts which had been translated directly into Arabic or indirectly through Syriac. Many of these Arabic translations were obtained by the Latin West from Moorish Spain. But also, after the collapse of Byzantium under Turkish pressure, eastern scholars migrated to the West taking earlier Greek manuscripts with them. Shortly after Ptolemy's magnum opus, known to the Arabs as *Almagest,* was translated into Latin from the Arabic version, his Greek *Syntaxis mathematike* found in Sicily was also translated into Latin.

Thus in the West in the late medieval period a great deal of Hellenistic astronomy with underlying Babylonian factual material, all passed through the alembic of Arabian astronomy, came to be widely known. The main acknowledgement of the Babylonian (spoken of as Chaldean) astronomers was of their eclipse records reported by Ptolemy; there were in addition a number of other, only later acknowledged, values such as planetary periods, an eclipse cycle, a calendaric (the Metonic) cycle and so on. Much of this Babylonian underlay became apparent only after the decipherment of the Mesopotamian cuneiform clay tablets in the late nineteenth and the twentieth centuries AD.

The important point is that Indian astronomy based on Babylonian, post-Hipparchan, pre-Ptolemaic Hellenistic astronomy provided a stimulus or a bridge for the revival of Hellenistic astronomy, through Arabian astronomy, of scientific astronomy in the West. As we shall see, there were in late medieval times numerous translations and numerous interpretations of late Hellenistic astronomy. Before we analyse them we should examine the contributions made by the Arabian astronomers (see Chapter 9).

CHAPTER EIGHT

ASTRONOMY AND ASTROLOGY

*T*hough many modern astronomers are touchy on the issue, there is a long association between astronomy, earlier confined to the study of the movements and positions of the celestial bodies, and astrology, the interpretation of these movements and positions as either omens for or determinants of the course of terrestrial, including human, events. The early Babylonians, and probably before them the Sumerians, regarded various celestial events (disappearances and reappearances of planets before and after conjunction with the Sun, solar and lunar eclipses, other conjunctions, and the presence of the Sun and the Moon in the opposite sides of the sky) as having ominous significance. The omens read in Mesopotamia could be favourable or unfavourable and were related to the king or to the community as a whole but not specifically to private individuals. Devastating floods, abundant crops, the defeat of the king's army or its success, the death of the king or his recovery from illness were the sorts of things read in the celestial omens. We have seen examples of such celestial omens in the Venus tablets of Ammisaduga.

This was part of a system of divination practised in Mesopotamia. Other bases for such divination were the direction of flights of birds, the state of the internal organs, especially the liver, of sacrificial animals, and dreams (see Hooke 1963). There is no allegation in such judicial divination that the celestial or other events control or determine the matters of immediate human concern. While an eclipse, or birds flying at a given time of the day in an unusual direction, or the liver of a sacrificial animal being in an unusual state, may be regarded as signs that circumstances generally are amiss, it does not thereby involve an assertion that these ominous events control as distinct from predict human affairs.

Judicial astrology, which flourished in the second millennium BC and the first half of the first millennium BC and which possibly originated in the third millennium, began to be reported in the surviving records in the second half of the first millennium BC in Mesopotamia as concentrating more on the individual person and on the disposition of the celestial bodies at the time of the person's birth. Whether this was an indigenous development or an importation is difficult to decide. This later horoscopic or genethlialogical astrology in which the location of the seven planets at the hour of birth provided the clues to a person's subsequent destiny was possibly a Hellenistic Eygptian contribution but it may have been an earlier Babylonian development transported into Egypt (see Sachs 1952a). It attached special significance to the regions of the zodiac where each planet was deemed to exercise greatest or least influence and to the general character of the planet's influence, benign or malign. This notion that the celestial bodies exercised a controlling influence on human and other terrestrial affairs possibly arose from the obvious influence of the prevailing position of the Sun in the sky on the seasons and in turn on various agricultural phenomena such as the ripening of the crops and the reproductive activities of the flocks. The Moon may have been seen, mistakenly, to have a part in this. The notion could then spill over to the five star-like planets.

In this developed genethlialogical astrology, it was important to be able to say what the disposition of the planets was at any hour at some specified earlier date. Hence, accurate tables of planetary positions within signs were of great importance (see Pingree 1963, 1974a).

Hipparchos is said to have written on astrology but no work of his in this field has survived. We do, however, have many slightly variant manuscripts of an astrological text by Ptolemy (second century AD), *The Mathematical Composition in Four Books*, frequently called *Tetrabiblos*. It is a quite lengthy work though not as long as his major astronomical treatise. Whereas the latter has a great deal of mathematical analysis and quantitative data, *Tetrabiblos* is qualitative and the argument entirely verbal. It is written authoritatively rather than apparently empirically and as a rule it seems to be stating a long-established body of knowledge or belief rather than providing an original contribution to knowledge. Ptolemy claims that the Sun, which is masculine, dry and hot, obviously controls the seasons and thereby plant and animal life; some influence is said to be complementary and exercised by the Moon, which is feminine, moist and warm. Therefore one would expect influences to be exercised by the other planets which are masculine or feminine, dry or moist, hot or cold and beneficent (Jupiter, Venus and the Moon) or maleficent (Saturn and Mars) or variable (Mercury). The masculinity–femininity of the planets is also subject to some variation – when in the east their femininity increases, whereas in the west their masculinity increases; that is, when they rise in the morning they are more feminine than usual and when they set in the evening they are more masculine.

The zodiacal constellations or signs were classified in several ways. First, the equinoctial and solstitial signs, *Aries* and *Libra*, *Cancer* and *Capricornus*, were warm and cool or hot and cold respectively and so are the relevant seasons they introduce; the seasons are 'solid' when the next signs follow an equinoctial or solstitial sign and manifest its quality more strongly; they are 'bicorporeal' when they follow a 'solid' sign and bring the relevant season to an end. There is an oscillation here in the significance of the zodiacal periods. Second, the signs were masculine and diurnal in the case of *Aries, Gemini, Leo, Libra, Sagittarius* and *Aquarius*, or feminine and nocturnal in the case of *Taurus, Cancer, Virgo, Scorpio, Capricornus* and *Pisces*. Third, they were in commanding and obeying pairs – *Taurus* and *Pisces, Gemini* and *Aquarius, Cancer* and *Capricornus, Leo* and *Sagittarius, Virgo* and *Scorpio*; or in pairs of equal power – *Gemini* and *Leo, Taurus* and *Virgo, Aries* and *Libra, Pisces* and *Scorpio, Aquarius* and *Sagittarius* (in each pair the relative length of day and night is the same and each member of the pair rises and sets on the same points on the horizon).

The Sun and the Moon each have their own 'house' – *Leo* and *Cancer* respectively, in which they exercise major influence. Each of the other planets has two 'houses' of maximum influence, for example Saturn has *Capricornus* and *Aquarius*, Jupiter has *Sagittarius* and *Pisces*, Mars has *Scorpio* and *Aries*, and so on.

Attention was paid to the 'aspects' of pairs of signs, for example a pair being in opposition when separated by 180°, or in 'trine' when separated by 120°, or in 'quartile' when separated by 90°, or in 'sextile' when separated by 60°. The signs may also be arranged in four sets of triangles, for example *Aries*, *Leo* and *Sagittarius*, or *Taurus*, *Virgo* and *Capricornus*. Each triangle is made up of either masculine or feminine signs and is governed by two planets. Each planet has a sign in which it exercises its maximum influence ('exaltation') or its minimum influence ('depression'). Thus the Sun has its 'exaltation' in *Aries* and its 'depression' in *Libra* whereas Saturn reverses this situation.

Depending upon the signs or thirds of signs (which were later called decans) in which two or more planets were located, the planets might reinforce or counteract one another's influences.

All of this is asserted dogmatically by Ptolemy as received 'knowledge' without any attempt to produce evidence even of an anecdotal sort.

In general (or judicial) astrology, one needed to determine the place or region, the time and the nature of the event to be predicted (earthquake, pestilence, flood, drought, conquest, etc.). Ptolemy divided the known surface of the Earth into four quarters – a series of 'climes' running south to north and a series of what we would call longitudes running east to west. Peoples in each quarter or segment of it came under the influence of the stars highest in their sky, thus the Gauls, Germans, Italians, Spaniards and Sicilians, being principally under the triangle *Aries-Leo-Sagittarius*, and so governed by Jupiter and Mars when those planets were in the west, were independent, liberty loving, fond of arms, industrious, possessing leadership and magnanimity and lacking passion for women. Whereas the Indians, Parthians, Medes, Babylonians and Assyrians, being principally under the triangle *Taurus-Virgo-Capricornus*, and so governed by Venus, are ardent, concupiscent, fond of adornment, dancing and luxury, and effeminate in dress. The place of some predicted event would be inhabited by peoples under influences of the sort producing the event. To the modern eye none of this is very persuasive – apart from the heat and drying power of the Sun, the properties of the planets are assigned on an arbitrary usually symmetrical scheme as are those of the zodiacal signs and no evidence is ever produced to validate any prediction.

Just as Ptolemy in *Tetrabiblos* asserted a special association or line of influence between certain planets in certain signs, and peoples in certain regions, he also asserted a special association between the former and certain regions of the human body, for example *Aries* with the head, *Taurus* with the neck, *Gemini* with the arms and so on. These associations were later stressed in the application of astrology to medical practice.

Ptolemy devoted at least equal attention to genethlialogical astrology, in which the bodily and mental features, length of life, manner of death, possible disease, quality of marriage and of offspring are predicted from the disposition of the planets in the signs at the time of birth of an individual. He says that the disposition of the planets at the time of conception would be preferable, but as the date of conception is usually not determinable, time of birth has to be used. He warns against the inaccuracy of sundials and water-clocks in determining time of birth and commends the use of an astrolabe. He cites many examples of the effects of joint influences of two or more planets located in specified parts of the zodiac but he emphasizes that the individual case is more complex than any example he cites, such as, 'Allied with Venus in honourable positions Saturn makes his subjects haters of women, lovers of antiquity, solitary, unpleasant to meet, unambitious, hating the beautiful, envious'.

In *Tetrabiblos* Ptolemy, as already remarked, seems to be reporting rather than constructing, and to be setting out a basic structure on which later Indian and Arabian astrology was built. Each had added features which will be dealt with only in general terms. At about the beginning of the Christian era, horoscopic astrology spread widely and rapidly, especially in the peripheral regions of the Roman world. Eastern mystery religions, including Christianity, were spreading in the same regions. The planetary weekday names, which were almost certainly astrological, indicate an earlier pagan Graeco-Roman association. Earlier Roman intellectuals such as Cicero rejected the claims of astrology and later the early Christian Fathers felt the need to combat it or to delimit it. The latter were opposed to it partly because it was so strongly associated with paganism and partly because its assumption of external determination of human affairs was at odds with the Christian doctrine of free will. The acceptance of the astrological weekdays by Constantine in the early fourth century AD must have been embarrassing though the pantheon of pagan gods could be accommodated by treating them as demons to whom there is no limit. Further, the problem of astral determination versus free will could be side-stepped by confining the former to material and non-human animate affairs which of course have a bearing on human affairs.

There are hints in *Tetrabiblos* of the use of genethlialogical astrology to establish occasions propitious for certain voluntary actions. These came to include medical treatments, business negotiations, marriages and so on. These applications became strong in later Indian astrology.

Later Greek and Roman writers, pagan, Christian and adherents of other mystical religions, began to support horoscopic astrology and to elaborate justifications of it. They also worked out in greater detail the steps in casting horoscopes and formulated apologies for occasions when the process seemed to break down. Defective predictions were attributed to (i) incompetent poseurs claiming to be skilled astrologers, (ii) inadequate information on the data, including hour and minute within the day, of the person's birth, and (iii) errors in the tables giving planetary positions in the signs of the zodiac and the three decans or houses within them at specified times; in this last respect the astrologers were asserting inadequacies of the astronomers.

Astrology spread with astronomy from the Graeco-Roman world into India at about the beginning of the Christian era and from there was later spread by Buddhist monks into China and South-East Asia. In the course of time the Indian astrologers added many celestial events to those specified by Ptolemy. For example they added the locations of the planets relative to the *nakshatra* stars to their locations in the signs of the zodiac and further introduced subdivisions such as *horas* ($15°$), *saptamsas* ($4\frac{2}{7}°$) and *navamsas* ($3\frac{1}{3}°$). They also took into closer account more exact planetary periods.

In India the association of astrology and astronomy has remained very close. *The Nautical Almanac*, for some time now a joint American-British production, was originally intended for both astronomers and navigators; more recently the needs of the former are catered for by a more detailed reference work, *The Ephemeris*. *The Indian Nautical Almanac*, a comparable compilation, has some material diligently computed by Indian astronomers of use only to the casters of horoscopes.

In the Sassanian Empire, founded in Iran in the third century AD, astrology was adopted from both Greek and Indian sources. It has some details of Greek astrology not adopted in Indian astrology and a few of the peculiar Indian additions.

Islam adopted astrology from three main sources – Greek, Indian and Sassanian. Many of the eighth and ninth century Arab scholars were more concerned with astrology than with basic astronomy. Amongst these were Masha'allah and Abu Ma'shar. As the superiority of Greek mathematical methods and data (however defective the latter) became evident Arabian astronomers and astrologers based their approach more and more on the Greek model, which they tried to improve, than on the Indian model. With the passage of the centuries the emphasis on astrology decreased in Islam. Nevertheless when astronomy moved, mainly from Spain into the medieval Latin Europe, astrology accompanied it. The Toledan and the Alphonsine tables were as important in western Europe for the casting of horoscopes as for purely astronomical purposes. Chaucer, an English public official and poet of the late fourteenth century, provides an interesting example of the concern with astronomy and astrology derived primarily from Islamic Spain. He refers to the Toledan tables 'ful wel corrected' (probably the Alphonsine tables corrected for longitudes and latitudes other than those of Toledo). He frequently characterizes a situation by referring to the astral influences at work at the time, or a person's temperament by referring to his or her natal planetary dispositions. He frequently dates some event by reference to some celestial phenomenon: this dating may be the identification of a year, a point within the year or a point within the month or within the day, but his specification often required a degree of technical knowledge beyond that of most of his contemporary readers. He wrote a treatise on the astrolabe based on a translation of an account by Masha'allah, and from the examples he gives of the use of the instrument it is clear he was competent with it. If a recently rediscovered manuscript, *The Equatorie of the Planetis*, is correctly attributed to him (Price 1955) he was familiar with another astronomical device and made a few observations of stellar altitudes from London. Chaucer was not, of course, either a professional astronomer or a professional astrologer. He was a talented amateur.

The close association between astronomy and astrology continued until the early seventeenth century. For a substantial part of his later career, Kepler (1571–1630) was employed as a court *mathematicus* or astrologer. Though a person of mystical inclination, Kepler almost certainly gave higher priority to his astronomical concerns. He seems not to have proposed any innovations in astrology whereas he made many proposals, a few futile but most of them effective, in astronomy.

CHAPTER NINE

ARABIAN ASTRONOMY
AND ITS IMPORTANCE
FOR THE REVIVAL
IN THE WEST

It is convenient to speak of the astronomy which was cultivated throughout the Islamic world from, say, the ninth to the fifteenth century AD as Arabian. The contributors to it were not all Arabs nor were they all Muslim. They worked in communities as widely spread as Samarkand (in modern Uzbekistan), Balk (in modern Afghanistan), Maragha (in modern northwest Iran), Baghdad (in modern Iraq), Damascus (in modern Syria), Cairo (in modern Egypt), and Toledo (in modern Spain). Though these contributors wrote mostly in Arabic, some wrote in Persian and a few in Hebrew. Ethnically they were mainly Arab but they also included Persians, Mongols, Syrians, Jews, Egyptians and Moors, and by religion they were mainly Muslims but they also included some Zoroastrians in Persia, Sabians (star-worshippers, in Harran, Mesopotamia), Jews and Christians, the last including, as already reported, such dissident sects as Nestorians and Monophysites in Mesopotamia and Persia, where they had taken refuge from the Greek Orthodox Church (see O'Leary 1949).

Arabian astronomy largely followed Ptolemaic astronomy. It concentrated on Arabic translations of Ptolemy's *Syntaxis mathematike* (called *Almagest*, 'the greatest') and commentaries on it. It made only a few theoretical improvements and put forward some notions which can only be regarded as retrogressions. It did, however, improve some Ptolemaic empirical values. These improvements included (i) the value of the obliquity of the ecliptic, (ii) the rate of the precession of the equinoxes, (iii) the length of the tropical year, and (iv) the direction of the solar apogee measured as an angle from the First Point of Aries.

Some of the Arabian astronomers thought that they were doing no more than improving Ptolemy's values, whereas some of them recognized that the discrepancies they discovered were the result of secular changes, that is, changes over time in the values.

Through new observations and/or calculations based on original or modified Ptolemaic theories, the Arabian astronomers brought up to date the Ptolemaic planetary tables and the Ptolemaic star catalogue. These were operations that needed to be repeated from time to time, partly as a result of errors in earlier observations and in the theories used to extrapolate from them and partly as a result of secular changes.

Among the retrogressions was the theory of trepidation which had the equinoxes increase for a time and then decrease. The several proponents of the theory of trepidation selected different angular oscillations. The proponents of the theory also disagreed on the period of the oscillation. They attributed the theory to unspecified Greeks, though we have only the vaguest record of such Greek thought.

It was thought for a long time that the melange of Aristotelian concentric spheres and crudely stated Ptolemaic deferents and epicycles to be found in Sacrobosco and other late medieval exponents of astronomy was a confusion of two diverse sets of conceptions introduced by some Arabian theorists. Until recently only an obviously incomplete version of Ptolemy's *Planetary Hypotheses* was available. We now have an Arabic version which indicates that Ptolemy himself produced the amalgam. It seems that Ptolemy accepted a thick sphere for each planet (between its perigee and apogee) which if given a physical interpretation, as Arabian thinkers often did, could be regarded as rolling around enclosed spheres in the 'ball-bearings' provided by the relevant epicycles.

The Ptolemaic planetary theories as presented in *Syntaxis mathematike* may be regarded as mathematical models valuable for predication of planetary positions but not necessarily representing a physical mechanism. Some Arabian theorists, as did some later European thinkers, made this distinction and were not always confident about how to resolve the dilemma.

Some attempts to improve the details of Ptolemaic geometrical planetary theories were made in the thirteenth and fourteenth centuries AD by al-Tusi, al-Shirazi and al-Shatir. Their attempts were aimed at removing the departures from uniform circular motion in Ptolemy's excentric deferents involving equants and crank-mechanisms. Only al-Shatir was reasonably successful. Interestingly, Copernicus used al-Shatir's geocentric version of the al-Tusi couple in a heliocentric setting. All this will be dealt with later.

Arabian astronomy happily accepted a number of Hellenistic notions which were rejected by the Church Fathers by about the middle of the first millennium AD. For instance, the account of creation in Genesis suggests a universe in flat layers rather than in concentric spheres. The Arabians had no difficulty in accepting that a solar eclipse was the result of the Moon blocking light from the Sun or that a lunar eclipse was the result of the Moon passing into the Earth's shadow. Some early Christian Fathers regarded such views and explanations as nonsense, if not blasphemous.

Rather than discuss the astronomical views of one Arabian astronomer after another it would seem better to have a chronological roll-call with a brief reference to personal contributions and a later survey of the topics one by one dealt with by these astronomers. The best accounts of the work of each are to be found in *The Dictionary of Scientific Biography* (Gillespie 1970–4). I have followed its Roman transliterations of the Arabic names. Nasr's two books (1968, 1976) provide valuable illustrative information and articles by Kennedy (1966) and his associate Roberts (1959) provide more technical information. It seems to me that there may be a great deal of Arabian astronomical material which has not been located, or if located not adequately interpreted; these are only hunches based on the remarks of the scholars looking into these matters.

———————— NINTH CENTURY ————————

Masha'allah (762–c. 815) was mainly a writer on astrology. He wrote an account of the astrolabe, a translation of which was the basis of Chaucer's *Treatise of the Astrolabe* (c. 1391). He also wrote an account of the armillary sphere described earlier by Ptolemy.

Muhammad ibn Musa al-Khwarizmi (c. 800–47), mainly a mathematician (our word 'algorithm' comes from the final element in his name), produced a set of planetary tables for the longitude of Ujjain, the early analogue in India of Greenwich. These tables were derived from the Indian astronomical treatises called *siddhanta* which were produced on several occasions between about the fourth and the seventh century AD. As stated above, this Indian astronomy was influenced by post-Hipparchan but pre-Ptolemaic astronomy. It made use of deferent-epicycle couples but lacked the detail and the complexity of the Ptolemaic analyses. It adopted the signs of the zodiac and numerous Babylonian quantities which had been adopted by Hipparchos.

Ahmad ibn Muhammad al-Farghani (died after 861), who worked in Cairo, produced an influential summary of the Arabic translation of Ptolemy's *Syntaxis*. Part of this summary was later translated into Latin and was widely used in Europe. Al-Farghani also wrote on the sundial and on the astrolabe.

Abu Ma'shar al-Balki (787–886), though primarily an astrologer, wrote an astronomical work using Indian methods and parameters. In his astrological writings he was strongly influenced by the Hermetic thought which had a Hellenistic Egyptian mystical origin and which was still current in Harran. He also manifests some corrupt Aristotelian influences. Though he came from Balk, he worked mainly in Baghdad.

Thabit ibn Qurra al-Sabi (836–901) was descended from the star-worshipping Sabian sect of Harran, where he did most of his work. He wrote several astronomical works in which he discussed the apparent motion of the Sun and the Moon, the visibility of the new Moon, and expounded the alleged trepidation or oscillation of the equinoxes, attributing the effect to the ninth celestial sphere. He also claimed an oscillation of the obliquity of the ecliptic.

———————— TENTH CENTURY ————————

Abu'Abd Allah Muhammad ibn Jabir al-Battani (died 928), known later in the West as Albategnius, was perhaps the most outstanding Arabian astronomer. He worked mainly in Raqqa, Mesopotamia. He developed relevant mathematical methods, for instance substituting the more modern sines for Ptolemy's (probably originally Hipparchos's) chords. He was an active and apparently accurate observer, producing a more accurate estimate of the tropical year than that adopted by Ptolemy and discovering the increase in the longitude of the solar apogee (in geocentric terms). Had Ptolemy as good an estimate of the positions amongst the fixed stars of the equinoxes and solstices as al-Battani had, he would have been able to produce a better estimate of the rate of precession. Al-Battani also produced a star catalogue which updated Ptolemy's longitudes for the calculated rate of precession in the intervening period (not correcting, however, Ptolemy's systematic mean error of 1° in his longitudes).

Muhammad Abu'l Nafa'al-Buzjani (940–97 or 8) produced a simplified version of *Almagest*. For a period of the nineteenth century he was mistakenly thought to have discovered the third lunar anomaly known as variation, in which the Moon at the octants (45° before and after conjunction and oppositions) is either ahead or behind its predicted mean position. He made important contributions to trigonometry and wrote valuable practical manuals on arithmetic and geometry.

'Abd al-Rahman ibn 'Umar al-Razi al-Sufi (903–86) wrote on the astrolabe, on the use of celestial globes and on astrology. He critically revised Ptolemy's star catalogue, correcting the longitudes by 12°42' (thus assuming 1° of precession in 66 years). From his own observations he pointed to some positional errors in Ptolemy's catalogue as corrected for precession. He checked the magnitudes of the stars, checked their colours, and identified Ptolemy's stars with Arabic names (e.g. 'first in the Ram's horn'). He seems to be the first to consider that what we now call the galaxy in Andromeda was not an ordinary star but a remote collection of stars.

Abu'l-Hasan 'Ali ibn 'Abd al-Rahman ibn Ahmad ibn Yunus (died 1000), working in Cairo, contributed many positional observations and produced a new set of planetary tables based on Ptolemaic theories. He reported the dates of many conjunctions of the planets with one another and with Regulus, of lunar eclipses and of equinoxes. He measured the obliquity of the ecliptic and the maximum latitude of the Moon. From Hipparchos's and his own measurements of the longitude of Regulus he produced a value for precession of 1° in 70.25 Persian years of 365 days, that is, in 70.2013 Gregorian years. The modern value is 1° in 71.713 Gregorian years and there is no evidence that it is changing.

——————— ELEVENTH CENTURY ———————

Abu 'Ali al-Hasan ibn al-Hasan ibn al-Haytham al-Besri (965–1040) made contributions to mathematics and optics as well as to astronomy. He claimed that all stars and planets, except the Moon, were self-luminous. He accepted the Ptolemaic theories using deferents and epicycles as no more than mathematical analyses useful for prediction but in place of them accepted the Aristotelian concentric spheres as a better account of physical reality. He deplored Ptolemy's resort to equants and objected to Ptolemy's fifth motion of the Moon. He wrote a commentary on *Almagest* as well as a treatise on optics.

Abu Rayhan Muhammad ibn Ahmad al-Biruni (973–1050), who worked in Khwarazm (modern Turkestan) and Iran, made many observations of solar altitudes in order to determine the dates of solstices and equinoxes and of eclipses. He wrote accounts of the construction and use of the astrolabe and the sextant. He wrote on astrology, on the chronological systems of various peoples and on the customs, the knowledge and social systems of India. He determined the terrestrial longitude and latitude of many places including eleven in India.

Abu-Ishaq Ibrahim ibn Yahya al-Zargali (*c.* 1028–87) began as an instrument maker in Toledo and with encouragement from the astronomers there he moved into astronomy. He accepted the concept of the trepidation of the equinoxes and estimated the motion of the solar apogee to be 1° in 299 years (thus separating it from precession). His version of the theory of Mercury implied that its orbit was oval. He assisted in the production of the Toledan tables which were widely used in Latin Europe as well as in Islam.

'Umar al-Khayyami (1048–131), better known in the West as Omar Khayyam, the Persian poet, mathematician and astronomer, is said to have helped reform the Persian solar calendar which had been based on the Egyptian calendar and which persisted alongside the Islamic lunar calendar. The Egyptian calendar consisted of twelve months of 30 days to which were added five epagomenal days. This calendar year was almost a quarter-day shorter than the tropical year so its first day moved forward in the seasons at a rate of about 0.2422 days per year or by a whole round of the seasons in about 1508 years. After a time the Persians coped with this anomaly by adding a thirteenth month to the calendar year after 120 years. This produced an average calendar year of 365.25 days, as in the Julian and Alexandrian calendars. In AD 1079 'Umar al-Khayyami proposed a remedy which made the tropical year on average less than 365.25 days. There are at least two sketchy accounts of what 'Umar al-Khayyami proposed and they do not agree in matters of detail (see the entry 'Djalali' in *The Encyclopaedia of Islam*). He seems to have proposed that usually a leap year of 366 days should occur after four years but now and then the leap year should occur on the fifth year rather than in the more usual fourth year. Qutb al-Din al-Shirazi, late thirteenth century, claims that this delay in the leap year occurred twice in a 70-year cycle, in the 29th and 70th year. This would result in an average calendar year of 365.24285 days, about 54.43 seconds too long. Ulugh Beg, fifteenth century, claims that this delay occurred once in 33 years, which results in an average calendar year of 365.242424 days, which is 14.12 seconds too long. There are puzzling irregularities in al-Shirazi's account of what 'Umar al-Khayyami proposed, so there is some reaon for preferring Ulugh Beg's account. It could be that neither is a good account of 'Umar al-Khayyami's proposal but that each is a later attempt to improve on it.

The Gregorian reform of the Julian calendar resulted in an average calendar year of 365.2425 days, which was 24.33 seconds too long in Gregory's epoch. 'Umar al-Khayyami's value, if we follow al-Shirazi, was inferior to the Gregorian value by about 30 seconds, or if we follow Ulugh Beg, superior by about 9 seconds – a remarkable achievement whichever account one takes.

──────── TWELFTH CENTURY ────────

Jabir ibn Aflah al-Ishbili (*fl.* in first half of twelfth century) wrote a general astronomical work which was later translated into Latin and widely used in Europe. He was very critical of Ptolemaic theories, largely on physical grounds. He failed to appreciate that Ptolemy's theories in *Almagest* were primarily mathematical models and not an account of what existed in nature. In his discussion he made little if any use of quantitative values.

Nur al-Din al-Bitruji (*fl. c.* 1190), following Jabir's attack on Ptolemaic theorizing, tried to work out a theory of concentric spheres on Aristotelian lines. He was unable to deduce from his theory any lunar or planetary positional values but believed this quest to be possible of accomplishment.

THIRTEENTH CENTURY

Muhammad ibn Muhammad ibn al-Hasan al-Tusi (1201–74), working in the observatory at Maragha in northwest Persia which was established under the patronage of Hulagu il Khan, grandson of Genghis Khan, improved the available sighting aids by increasing their size, for example he used a quadrant with a 3 metre radius. After a dozen years of observation he and his assistants produced a new set of planetary tables, the Ilkanic tables. Al-Tusi also tried to avoid, as reported above, the Ptolemaic departures from uniform circular motion as involved in excentric deferents, equants and crank-mechanisms. These all implied a changing distance of the 'planetary' body from the Earth, as will be explained below. Al-Tusi applied his geometrical analyses to the planets without complete success for he had great difficulty in finding appropriate diameters for the component circles. Qutb al-Din al-Shirazi (1236–1311), at first at Maragha and later Tabriz, employed the analyses more effectively but on a more limited scale; even he, however, was not satisfied with his attempts.

FOURTEENTH CENTURY

'Ala' al-Din Abu'l Hasan 'Ali ibn Ibrahim ibn al-Shatir (1304–75), working in Damascus, succeeded in what al-Tusi and al-Shirazi had attempted without real success. For the Sun and the Moon he used two epicycles, the primary one rotating in the opposite sense to the deferent and the secondary epicycle, on whose circumference the celestial body was borne, revolving in the same sense as the deferent. For the star-like planets other than Mercury, al-Shatir used three epicycles, and for Mercury he added a fourth and a fifth epicycle to form an al-Tusi couple. Interestingly, Copernicus two centuries later used the al-Shatir constructions, with much the same ratios of radii. Al-Shatir's analyses, of course, were geocentric whereas Copernicus's were heliocentric.

―――――― FIFTEENTH CENTURY ――――――

Ulugh Beg (died 1420) was the grandson of the Mongol Tamarlane. He established an observatory in Samarkand with instruments on a larger scale than any used before. For example he built a sextant some 60 metres in radius aligned to the meridian. By means of such instruments he observed stellar positions and produced the first post-Hellenistic star catalogue based mainly on original positional observations rather than mainly on corrections of Ptolemy's catalogue for precession.

―――――― TOPICS OF ARABIAN ASTRONOMY ――――――

Among the empirical questions addressed quite early by the Arabian astronomers was the size of the Earth. On two separate occasions teams were sent out to measure the distance between sites differing by 1° of latitude. The values obtained were from 56 to 57 Arabian miles, presumably about 119 000 metres (a value somewhat too great).

Another instance of concern with an empirical issue was the establishment of the direction or longitude of the solar apogee, the Sun's maximum displacement from the Earth. This was worked out from the lengths of the seasons, a method employed by Hipparchos, who found the value to be 65°30' from the First Point in Aries. Ptolemy claimed to have confirmed this value though we know that it must have been larger by his time. Al-Battani, tenth century, and ibn Yunus and al-Zargali, both eleventh century, found values respectively of 82°17', 86°10' and 77°50'. All recognized that in respect of the longitude of the solar apogee allowance had to be made for the precession of equinoxes but the departures of their values from Ptolemy's alleged value could be accounted for only in part by this consideration. Perhaps only al-Zargali appreciated that the longitude of the solar apogee was changing. Al-Battani and ibn Yunus probably thought that they were correcting a Ptolemaic error. Ironically al-Zargali's was the poorest measurement of the three for the relevant epoch. The increase in the longitude of the solar apogee is about 19' per century, which al-Battani's values would have approximated had Ptolemy not misled him both about that longitude in, say, AD 135 and about the dates of the equinoxes at that same time.

The Arabian astronomers devoted great efforts to establishing the dates of equinoxes and solstices, which enabled them through comparisons with dates established earlier to estimate the length of the tropical year and the rate of the precession of the equinoxes. Ptolemy, as reported above, in giving the dates and times of three equinoxes which he alleged he had observed between AD 132 and 140, placed these events according to Newton (1977) about 28 hours later than modern considerations suggest. Hipparchos's estimation of the tropical year was about 0.0043 days (6 minutes 11.52 seconds) too long. If Ptolemy calculated, rather than observed, the dates of these equinoxes by multiplying Hipparchos's intervals by a value 0.0043 days too long, he would have arrived at dates about 28 hours too late. Al-Battani, using Ptolemy's claimed dates and times for these equinoxes and his own observations of equinoxes in AD 880, produced a value for the tropical year which was about two minutes of time too short for his epoch. Had he gone back to Hipparchos's dates and times he would have been in error by only a few seconds.

Again using Ptolemy's alleged dates and times of equinoxes, al-Battani estimated that precession had been occurring at the rate of 1° in 66 years since Ptolemy's epoch. Ibn Yunus, who wisely went back to Hipparchos's measurement of the longitude of the star Regulus, which he compared with a value he had established for his own epoch, concluded that the rate of precession was 1° in 70 years. These several values for the rate of precession led some Arabian astronomers to adopt a crudely formed pre-Ptolemaic Hellenistic view that the rate of precession oscillated or was subject to trepidation. One Arabian view was that stellar longitudes decreased by 8° in 640 years and then increased by 8° in the next 640 years. Another version of the theory of trepidation accepted the oscillation of ± 8° but set the period at 800 years. There is no currently acceptable evidence for any secular variation in the rate of the precession of the equinoxes.

Another empirical value in which the Arabian astronomers were interested was the obliquity of the ecliptic. Ptolemy claimed to have confirmed Eratosthenes's value of 23°51'20", a value a little too large in Eratosthenes's own epoch and still larger in Ptolemy's epoch (23°42' would have been nearer). During the tenth and eleventh centuries al-Battani, ibn Yunus and al-Zargali found values of 23°33', 23°35' and 23°33' respectively, all fairly good approximations for their epochs. Ulugh Beg found the value 23°30'17" (only 32" in error). The present value is 23°26'24" and is still decreasing, though gravitational considerations are said to suggest that sometime in the future it will begin to increase. Al-Zargali, perhaps on analogy with the hypothesis of trepidation in rate of precession, suggested that the obliquity of the ecliptic oscillated between 23°53′ and 23°33′; the latter alleged minimum, however, has been exceeded in our time.

Hipparchos, followed by Ptolemy, considered that the Moon varied in latitude from the ecliptic by about ± 5°. Several Arabian astronomers put forward limits ranging from ± 4°45′ to ± 5°8′. There are indeed variations in the limits but the precise values and the patterns seem to have eluded the Arabians. The lunar inclination varies over a mean range of ± 5°9′ in 18.61 tropical years with a minor oscillation of ± 9′ in a 174-day cycle laid on top of the main variation of up to ± 5°.

Several Arabian astronomers produced astronomical tables giving planetary positions at specified times; these positions were usually computed from the Ptolemaic planetary theories using contemporary observational values rather than Ptolemy's. A star catalogue was also usually included but until Ulugh Beg, fifteenth century, as reported above, the stellar longitudes in these catalogues were normally those of Ptolemy's catalogue corrected for whatever was taken to be the rate of precession during the intervening period but leaving Ptolemy's mean error of about − 1° longitude unaltered. The sets of tables also included terrestrial longitudes and latitudes of important towns; in the Toledo tables, for example, geographical locations for numerous centres from Tangier in present Morocco, 5°50′W, to Kabul in present Afghanistan, 69°10′E, were given, though the stated terrestrial longitudes are often hard to reconcile with modern values. This information was important for time-reckoning to establish prayer times and for establishing the direction of Mecca which had to be faced during prayers.

Ulugh Beg's star catalogue contained data on 1018 stars. He admitted that in a few cases he had reduced a longitude from Ptolemy's catalogue because the star was below his southern horizon or because he had observed the longitude of only one of a pair of neighbouring stars and had used the difference in longitude given by Ptolemy to infer the value for the other member of the pair. He copied the latitudes of many stars from Ptolemy's catalogue, or having observed the latitude of a star used

Ptolemy's differences, to calculate the latitudes of neighbouring stars. Knobel (1917) suggests that only about 700 stars in Ulugh Beg's catalogue had both longitude and latitude established by original observation. From data given by Peters and Knobel (1915) on Ptolemy's catalogue and by Knobel on Ulugh Beg's, the mean error in the latter, which is about −13′, is much less than that in the former, which is about +41′.

As has been reported, most Arabian astronomers adhered to the Ptolemaic geometrical planetary theories in order to calculate with the aid of current observational parameters the astronomical tables they needed. A few theorists such as Jabir ibn Aflah and al-Bitruji, both in the twelfth century, queried the Ptolemaic geometrical models on the grounds that they did not conform to physical reality as depicted in Aristotelian physics. They could not, however, work out the mathematical models required to predict planetary positions using the concentric Aristotelian spheres which they wished to emphasize.

Some other Arabian astronomers, as already mentioned, tried to propound geometrical analysis which by avoiding excentrics, equants and crank-mechanisms conformed to the so-called Platonic requirement that anomalous planetary motions be reduced to component uniform circular motions. Al-Tusi (1204–74) was the first to make the attempt. He introduced what Kennedy (1966) has called the 'Tusi couple', which generates a lengthening and shortening distance between an orbiting planet and some central reference point such as the Earth. The centre of a small circle was deemed to rotate anticlockwise in a circle of double its diameter; meanwhile the planet borne on the circum-ference of the smaller circle rotated clockwise. As a result the planet moved in a straight line away from and towards the central reference point (see Fig. 9.1). The couple, of course, is borne around a deferent or even a deferent-epicycle pair, operating on the principle of uniform circular motion relative to the centre. Al-Tusi could not find the required parameters (periods of rotation and relative lengths of radii) to enable the analysis to be effective in prediction. His younger colleague, al-Shirazi (1236–1311), was only slightly more effective.

Al-Shatir (1304–c. 1379) achieved what al-Tusi and al-Shirazi were attempting by introducing a second epicycle for the Sun and the Moon, a second and third epicycle for Venus, Mars, Jupiter and

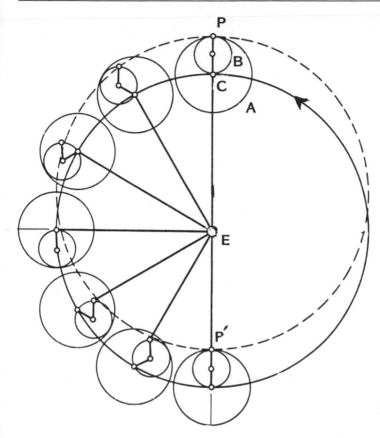

Fig. 9.1   *The 'Tusi couple' as used by al-Shatir in his theory of Mercury. The centre, C, of the circle, A, is borne anticlockwise on the circumference of the third epicycle (the deferent and the first and second epicycles are not shown in the diagram). Rolling anticlockwise inside circle A is circle B, the diameter of which is half that of A. The planet Mercury (P) borne clockwise on the circumference of P has its distance from Earth reduced by twice the radius B from P to P'.*

Saturn, and a 'Tusi couple' at the end of the radius of the third epicycle for Mercury.

For the Sun, al-Shatir, like al-Tusi, centred the deferent on the Earth and he considered the Sun to revolve west to east on a radius of 1,0;0 (60 in decimal terms) in a mean year. A first epicycle with a radius of 4;37 (4.617 in decimal terms) was deemed to revolve east to west so that its radius remained parallel to the Earth-Sun apsidal line joining the solar apogee and perigee. A second epicycle was centred on the circumference of the first epicycle. Its radius was 2;30 (2.5 in decimal terms) and revolved at double the speed of the Sun in the same direction as that of the first epicycle.

Al-Shatir's theory of the Moon was similar to his theory of the Sun but had different parameters for the lengths of the radii and for their velocities of revolution (see Roberts 1957). For Venus and the three superior planets, al-Shatir used a deferent and three epicycles (see Kennedy and Roberts 1959). The radii of the first and the second revolved in the opposite direction to the radius of the deferent, the radius of the first in the same period and that of the second epicycle in half the period of the deferent. The radius of the third epicycle revolved in the same direction as that of the deferent but with a velocity such that it remained parallel to the line joining the Earth and the Sun. For Mercury, al-Shatir located a 'Tusi couple' at the end of the radius of the third epicycle, thus in effect using five epicycles. The resultant path of Mercury was a somewhat distorted oval.

Just as for the Sun and the Moon, each planet had its distinctive relative lengths and periods of rotation of the several radii (see Table 9.1). Avoiding Ptolemaic excentrics, equants and 'crank-mechanisms', al-Shatir was able to provide predictive bases equally effective as those of Ptolemy.

Many Arabian astronomical writings were translations of, summaries of or commentaries on what they called Ptolemy's *Almagest*. The commentaries were meant to clear up obscurities, to provide improved empirical values or to draw attention to what were held to be deficiencies. Others consisted primarily of updated planetary tables, many based on Ptolemaic theories but with more recently established observational data bases, or stellar catalogues based on Ptolemy's catalogue corrected for, usually incorrect, rates of precession. Ulugh Beg's star catalogue was a partial exception. A few Arabian writings were general astronomical treatises meant for advanced students who might be preparing for independent work. Al-Tusi produced one such compendium and al-Shatir another.

The Arabian astronomers were probably the first to set up observatories in a formal sense equipped with various sighting and

Table 9.1 Relative lengths of radii in al-Shatir's planetary theories

	Relative lengths of radii			
	Deferent	*1st epicycle*	*2nd epicycle*	*3rd epicycle*
Mercury	1,0;0	4,5	0;50(?)	22;46
Venus	1,0;0	1;41	0;26	43;33
Mars	1,0;0	9;0	3;0	39;30
Jupiter	1,0;0	4;7,30	1;22,30	11;30
Saturn	1,0;0	5;7,30	1;42,30	6;30

These relativities hold within a planetary analysis but not between them; thus the radius of the deferent is set to 1,0;0 (60 in decimal terms) in each case although the planets from Mercury to Saturn were at increasing distances from the Earth, which was the centre of each of the deferents. The radii of Mercury's fourth and fifth epicycles were 0;33, for every deferent should not be taken to indicate that all deferents were the same size, and 0;16,30 respectively. Note these are sexagesimal numbers, thus 1,0;0 equals 60 in a decimal system and 4;5 equals $4 + \frac{5}{60}$ or 4.0833.

timing instruments and staffed by several astronomers including masters and pupils. Baghdad in modern Iraq may have been the site of the first of these observatories, followed by Maragha in northwest modern Iran, followed by Damascus in modern Syria and Cairo in Egypt, then Toledo in modern Spain and next Samarkand in modern Turkestan. The most popular instrument was the astrolabe which required a replaceable plate appropriate to the terrestrial latitude where it was being used. It was both a sighting and a calculating device. Another important instrument was the armillary sphere, a series of graduated metal circles offset at various angles; this was described by Ptolemy as an *astrolabon*, but it was quite different from the planispheric astrolabe which the Arabians, through their fine instrument-making, made famous. The Arabian astronomers following the Greeks also used quadrants and sundials which they continued to increase in size in order to achieve greater precision. The remains of Ulugh Beg's instrumentation in Samarkand give some indication of the size of these devices late in the piece. The well-preserved observatories set up by the Hindu prince, Jai Singh, in Delhi, Jaipur and elsewhere in the eighteenth century, though anachronisms, as the telescope was already well developed in the West, probably give a fairly true impression of the late Arabian observatories on which they were almost certainly modelled.

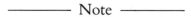

Note

Arab names tend to have special features. The Roman name 'Caius Julius Caesar' cites in order a personal (or given) name, a family name (or surname) and the name of the *gens* (or clan or tribe) to which the family belonged. The family name (*cognomen*) and the clan name (*agnomen*) descended patrilineally. In English usage we tend to have a surname inherited from the father and one or more individual given names. Sometimes the surname of a wife is prefixed to that of her husband to generate a novel surname, for example Smith-Young. In Spanish this is a standard practice; the son of a mother whose father's basic surname was Cortes and of a father whose basic surname was Hernandez may be called Rogelio (his given name) Cortes y Hernandez (his conjoint surname). The basic surname of both parents in this case is patrilineal.

Arab names tend to be genealogical, thus Muhammad ibn Muhammad ibn al-Hasan al-Tusi (Muhammad the son of Muhammad the son of al-Hasan from Tus). In addition to place-names such as al-Tusi, al-Shirazi and al-Balki at or near the end of a long string of Arabian names were other ancestral identifications, for example, Thabit ibn Qurra al-Sabi (Thabit the son of Qurra the Sabian, a star-worshipper). Further in a long Arabian name the honorific al-Din (perhaps the honorable) was added.

CHAPTER TEN

EARLY
ASTRONOMICAL
INSTRUMENTS

*E*arly astronomers, say from the second half of the first millennium BC, though earlier in one respect, wishing to establish astronomical quantities sought three categories of data. One was the apparent brightness of the fixed stars and the varying brightnesses of the star-like planets. A second was the specification of the time within the day or the month or the year when some celestial event occurred. Such events included planetary disappearances and reappearances before and after conjunctions with the Sun, eclipses and the like. Time intervals between such events were also important. A third was the establishment of (i) a celestial position or direction within some frame of reference, such as the ecliptic longitude and latitude favoured by the Babylonians and the Greeks or the altazimuth system often preferred by the Arabian astronomers emphasizing the horizon, or (ii) the angular separation of prominent neighbouring stars.

Until modern times, stellar magnitudes could be established only by subjective naked-eye judgements. Ptolemy, and probably before him Hipparchos, established six grades of stellar brightness magnitudes. We know that Ptolemy designated fifteen stars as being magnitude 1 and went down to stars of magnitude 6, which are just visible to naked-eye observation on a clear moonless night. Modern, post-Copernican aids, such as photometers, photographic plates and photo-electric cells, have shown that on average stars classified by Ptolemy one magnitude greater than dimmer stars deliver to the eye radiant energy 2.52 greater than stars one magnitude lower. In general as we move from one magnitude to the next below it, the number of stars increase two to three times. The numbers of Ptolemy's magnitude 1, 2 and 3 stars conform to this pattern, but the number he allotted to magnitude 4 did not increase over that in magnitude 3 by as high a proportion and the numbers in his magnitude 5 and 6 progressively decreased; he was not very successful in identifying lower magnitude stars.

Of the fifteen stars classified by Ptolemy as magnitude 1, one of these, Denebola, *beta Leonis*, has since been dropped to magnitude 2. Eight classified by Ptolemy as magnitude 2 have been raised to magnitude 1; seven classified by him as magnitude 1 have been recognized as one magnitude brighter, namely 0, and two, Sirius and Canopus, as two magnitudes brighter, namely −1. There may have been some changes in the brightness in these stars in the almost two millenniums since Ptolemy made his assessments, but it is more likely that the differences between the Ptolemaic and the modern classifications are the result of the introduction of more effective observational aids.

We know something about Babylonian time-reckoning aids but very little about their sighting devices for use in establishing celestial directions. Celestial events themselves were used extensively in Mesopotamia in time reckoning. The calendar year consisted of twelve or thirteen lunar synodic months (from one new crescent Moon after a sunset to the next) rounded to 29 or 30 days. An ordinary calendar year of twelve synodic months totalled 354 or 355 days in the ratio of about 2:1. Such calendar years are about ten or eleven days shorter than the year of the seasons or the similar sidereal year (intervals between successive conjunctions of the Sun and some fixed stars), which were approximated by the average intervals between successive heliacal risings of some fixed stars. Hence every now and then a thirteenth month was intercalated or added. From about 500 BC in Babylonia this intercalation was done by rule so that in nineteen calendar years there were seven 'swollen' (embolistic) years of thirteen months and twelve 'hollow' years of twelve months each. Thus in

Fig. 10.1 *An early Egyptian shadow-clock. The end with the crossbar was turned to the rising Sun and the shadow stepped out the five temporal hours until midday when the Sun was overhead. The device was then turned around so that the shadow could step out the five temporal hours until sunset.*

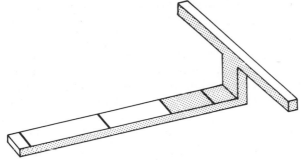

nineteen calendar years there were 235 synodic months which came to only about one-third of a day short of nineteen sidereal years. The uncertainty as to whether a future month would have 29 or 30 days was solved by using the so-called 'lunar day', a thirtieth of a month, later called by the Indians a *tithi*; it was a full solar day in a 30-day month but only 0.96 of a solar day in a 29-day month, a difference of a little less than an hour, which would have been of little consequence until recent times.

The absence of mechanical or electrical clocks provided a problem in respect of points of time within the day. We know that in ancient Egypt shadow devices (see Fig. 10.1), of which the sundial is a later example, were in use for subdividing the daily period of sunlight. Unfortunately the Egyptians complicated the system by wanting to have ten 'hours' of sunlight to which were added the two 'hours' of dawn and of dusk, plus twelve 'hours' of night (that is, while the stars were shining), irrespective of the season. At Cairo in Egypt, the ten 'hours' of sunlight could each be as short as about 60 minutes by our clocks in midwinter, about 84 minutes in midsummer and about 73 minutes in spring and autumn. At Cairo, the two twilights would be about 60 minutes. This leaves about twelve of our clock hours for night at midwinter but only about eight at midsummer. As each night had twelve of these 'temporal hours', the 'hours' at night were about 60 of our clock minutes in midwinter but only about 40 of our clock minutes in midsummer.

The Sun cannot be used for measuring the passage of the night. So the Egyptians selected a set of stars in a curving band south of the ecliptic, the members of which rose and set at approximately 40-minute intervals, possibly at shorter intervals in a summer than in a winter night. Some twelve of them rose and set during the course of a night (from the end of dusk to the beginning of dawn). Each one of them rose four minutes earlier night by night (as a result of the Sun's apparent eastward progress in a sidereal year), so after ten days each member in the series rose at the time its neighbour to the west had risen. Because of this successive replacement after ten days, they came to be called, following the Greeks, *decans*. There were ultimately 36 *decans* and Hellenistic Egyptian horoscopic astrology divided each sign of the zodiac into three decans each 10° in longitude. The influence of a planet on both general and individual human affairs was deemed to vary not only by its place within the signs but also by its place in decans within the signs.

Other devices had been used for measuring the passage of time during the night. Two of these could also be used during the daylight (sunlight plus twilight). One was the water-clock, or as the Greeks called it, *clepsydra*, the water-stealer. In its simplest form it was an earthenware or stone pot with a small hole in the bottom through which the water with which it was filled slowly leaked. On the inside walls of the pot were graduations. In a preserved early Egyptian model, the spacings of the graduations are different for each of the three 'seasons' to accommodate the changing 'temporal' hours.

The Babylonians also used the water-clock and as a rule they also used it to check the passage of twelve 'temporal hours' or some analogous periods such as six 'temporal double hours', which they called *berus*. They left many instructions on how much water in terms of weight with which to load a water-clock in summer and in winter. The ratio was 3:2 by weight of water for winter-night as compared with summer-night. This ratio is not quite accurate in terms of the relative lengths of winter-night and summer-night hours at the latitude of Babylon but it is not far wrong. There is, however, another consideration. The rate at

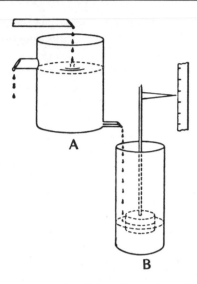

Fig. 10.2   *A constant head clepsydra such as devised by Ktesibios. Vessel A has a constant volume of water as any surplus fed into it is released through its upper outlet. With a constant volume of water it releases water at a constant rate through its lower outlet into vessel B.*

Reproduced with permission from C. Singer *et al.* (eds), *A History of Technology*, Vol. 3, Oxford, Clarendon Press 1957.

which water leaks out of a pot with an exit of given diameter depends on the weight of water remaining in the pot if its sides are parallel. A pot with parallel sides leaks more slowly as the water level drops. The Egyptians tried to overcome this by tapering the sides of the pot inwards by 70°. Ktesibios, an Alexandrian Greek of the third century BC, solved this problem. He used a series of three vessels. The first, the basic reservoir, fed slowly into the second, the water level of which was kept constant by leakage through a hole in its side. The second vessel uniformly trickled water into the third, a rising float in which measured the passage of time (see Fig. 10.2).

In the first century AD two other timing devices were being used: the hourglass in which sand dropped from an upper chamber through a narrow passageway into a lower chamber, and the 'burning clock'. The latter took several forms, such as a graduated candle which burned its way down through the hours and minutes (four hours was about the maximum space) or a lighted wick fed by a bowl of oil the height of which was reduced with the passage of time. In all forms of 'burning clocks' it was important to shield the flame from varying currents of air. It is not clear whether hourglasses or 'burning clocks' were ever put to astronomical use. The Chinese from perhaps earlier times had an ingenious 'burning clock'. Slightly inflammable material was placed in channels which were placed at right angles to each other and were carved in a flat stone or impressed into a ceramic block. When the material was lit its burning away along one straight channel marked some fraction of time.

Mechanical clocks were developed largely in the second millennium AD. At first they were gravity activated by a falling weight, later they were spring activated and later still electrically activated. A long-standing problem resided in ensuring a constant or near constant rotation of the gears controlling the hour, the minute and later the second hands; that is, in providing a governing device to keep within limits the rate of the delivery of energy by the motivating device. An early clock governor was the foliot balance control; it was very inaccurate, allowing an error of half an hour per day. A later device was the pendulum controlled escapement wheel and a still later one the hairspring controlled escapement wheel. As these are largely outside the period of astronomy being considered, there seems little point in going into details about them. There are numerous informative accounts of the development of the mechanical clock and its successors (see especially, Landes 1983).

We have virtually no information on Babylonian sighting and angle-measuring devices, though we can be confident that they had them. The Babylonians or possibly the Sumerians divided the circle into 360 degrees, perhaps at a time when they believed the year to be 360 days. Each degree was divided into 60 minutes (our word comes from the Latin *pars minutae*, the small part or fraction) and each minute into 60 seconds (from *pars secundae*, second fraction). This is all part of the Mesopotamian sexagesimal number system. In our decimal system we could reasonably describe 0.1 to 0.9 as our first fractions, 0.01 to 0.09 as our second fractions and so on. As has already been asserted, it should not be inferred that the

Babylonians could measure positions or angular differences to an accuracy of a minute, much less a second or a 'third' of arc; their very precise positional values seem to be the product of calculations from less precise observations. These basic observations must have used sighting devices which enabled directions or differences to be established to a degree and substantial fraction of a degree, possibly $\frac{1}{3}$ of a degree. But we have no good indication of what these devices may have been. Reference is made to a pole or gnomon which may have been used to establish celestial altitudes and directions. It would be surprising, however, if some of the Greek instruments of the later part of the first millennium BC were not in use in Babylonia, at least in the Seleucid era (from 311 BC) and possibly a little earlier.

The availability of information about sighting instruments is different for the Greeks and the Arabs. From at least the time of Hipparchos, second century BC, several types of sighting devices were in use by the Greeks for the establishment of directions or differences in direction. It would seem that Eratosthenes, in establishing the obliquity of the ecliptic, measured the altitudes of the Sun at the summer and the winter solstices and halved the difference. It is reported that he found the difference between solstitial altitudes to be $\frac{11}{83}$ parts of a circle (which yields an obliquity of 23°51′18″41‴). This reported finding suggests that whether Eratosthenes established the solstitial difference by means of a quadrant or an armilla or some other device, the device may not have been graduated in 360° for the circle, 60′ for the degree and so on. Yet a division of a circle or a semicircle or a quadrant into 83 parts would not be mechanically easy to achieve. Perhaps Eratosthenes's $\frac{11}{83}$ was a value interpolated between two round numbers, such as $\frac{6}{50}$ and $\frac{7}{50}$ or $\frac{7}{60}$ and $\frac{8}{60}$.

In estimating the circumference of the Earth Eratosthenes used the following data: at the summer solstice the Sun at midday was directly overhead at Syene but was $\frac{1}{50}$ of a full circle from the zenith at Alexandria, and Alexandria was 5000 stadia north of Syene. Hence the circumference of the Earth was deduced to be 250 000 'stadia'. There are several values of a 'stadium' to choose amongst; one of them yields a good approximation to the presently accepted circumference of the Earth. There are two accounts of the observation at Syene. In one, a collaborator of Eratosthenes is said to have been at the bottom of a deep well and to have been able to see the Sun overhead on the solstice. In the other it is said that it is known that at Syene a vertical gnomon cast no shadow at midday on the solstice. It is said that, at Alexandria, Eratosthenes observed that at midday on the solstice a vertical gnomon cast a shadow in a hemispherical bowl, as stated above, at $\frac{1}{50}$ of a full circle. This suggests a different graduation from that on whatever device Eratosthenes used to estimate the obliquity of the ecliptic. It could be, of course, as already suggested, that $\frac{11}{83}$ and $\frac{1}{50}$ are interpolations within scales, not fractions of 83 or of 50 but of some other range of values; it is worth recalling that Archimedes gave the value of *pi* as greater than $3\frac{10}{71}$ but less than $3\frac{1}{7}$, in decimal terms greater than 3.140845 but less than 3.1428571.

A very crude angular observing device was called Archimedes's staff, and in late medieval and early modern times, Jacob's staff. It consisted of a graduated staff, one end of which was held closely in front of the eye. Along this staff a cylinder (in Archimedes's form) or a crossbar or a disc was slid forwards until the distance between the two points, say between two stars, was just spanned; the angle subtended by the variable piece could be read from the graduations on the staff. Archimedes is said to have measured the angular diameter of the Sun by means of his staff. Later mariners used the Jacob's staff to establish the altitude of celestial objects crossing the meridian.

The quadrant may have been inspired by the sundial. It consisted of a square block with a shadow-casting perpendicular bar or gnomon. A 90° quadrant was centred on this gnomon and graduated in degrees. The instrument was aligned in the meridian plane and the altitude of the Sun and the Moon, when bright enough to cast a shadow, could be read on the graduated quadrant. Later, by using a movable arm with sights the instrument could be used to measure the altitude of stars or the separation of pairs of celestial bodies.

Another sighting device, invented perhaps as early as the third or fourth century BC, was the diopter. In its developed form it consisted of a rotatable arm with a backsight at one end and a foresight at the other. This arm could be swung around a graduated tiltable plate. When a celestial object was captured in the sights, the tilt of the plate gave the object's altitude and the direction of the arm gave its azimuth.

Another device known as a triquetrum or *regula Ptolemaica* was

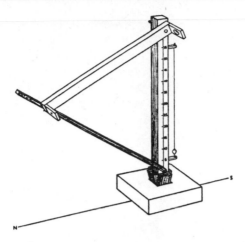

Fig. 10.3 *The triquetrum or* regula Ptolemaica. *The star was sighted through the holes at the ends of the upper adjustable arm, the angle of which could be read from the lower arm.*

Reproduced with permission from C. Singer *et al.* (eds), *A History of Technology*, Vol. 3, Oxford, Clarendon Press 1957.

probably invented by Hipparchos. A vertical post about 1.8 m long had two movable arms, one hinged near the top and the other near the bottom of the post. The device was so oriented that these two arms could move only in the plane of the meridian. The upper arm had a frontsight and a backsight through which a celestial object could be viewed. The lower end of this arm slid along the lower arm and pulled it up or down. The lower arm was graduated and the altitude of a celestial object seen through the sights could be read on a 0 to 60 scale (see Fig. 10.3).

Several Hellenistic instruments came to be called *armillae* from the Latin *armilla*, a bracelet. They consisted of one or more metal rings of rectangular cross-section. The simplest was the equatorial armilla, a ring often about 400 mm in diameter which was fixed in the equatorial plane. At the equinoxes the upper part of the ring cast a shadow on the lower part and so marked the time of the equinox. A more elaborate version consisted of an outer ring set in the meridian plane and an inner ring which could be slid in the same plane as the outer one. The inner ring had two markers, one

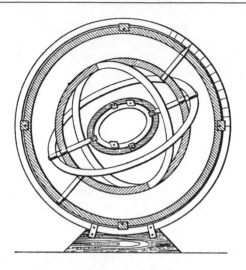

Fig. 10.4 *The armillary sphere. The star was sighted
through the holes in the tabs on the innermost ring which could
be rotated within the next ring. The pair could be tilted within
the next set of rings. The outermost ring was kept on the
meridian and the ring within it could be rotated in order to give
a value of altitude.*

Reproduced with permission from C. Singer *et al.* (eds), *A
History of Technology*, Vol. 3, Oxford, Clarendon Press 1957.

to cast a shadow and the other to receive it. This enabled, by
reading from the graduation on the outer ring, the assessment of
the altitude of the Sun at midday on any occasion.

From quite early, say from Hipparchos in the second century
BC, armillae seem to have been graduated in terms of 360° for the
full cycle and some fraction, not necessarily $\frac{1}{60}$ of a degree; indeed
$\frac{1}{2}, \frac{1}{3}, \frac{1}{4}$ and $\frac{1}{6}$ seem to have been more likely fractions of a degree
in use.

A much more complex armilla later called the armillary sphere
was in use probably as early as the second century BC (see Fig. 10.4).
Ptolemy gave a detailed description of such an instrument though
it may have been devised by Hipparchos. It enabled amongst other
things the establishment of the ecliptic longitude and the latitude
of any celestial body at a given time. It is difficult not to believe

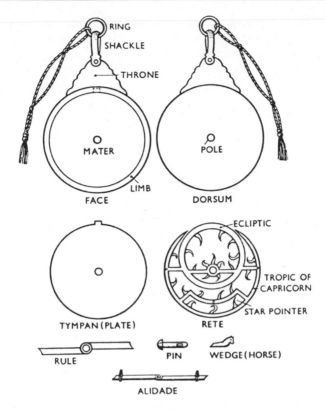

Fig. 10.5   *The components of a typical Arabian planispheric astrolabe.*

Reproduced with permission from C. Singer *et al.* (eds), *A History of Technology*, Vol. 3, Oxford, Clarendon Press 1957.

that the Hellenized Babylonians had a device of this sort. The armillary sphere was used extensively in Hellenistic, Arabian and late medieval Chinese and Western astronomy.

Another device almost certainly of Hellenistic origin but brought to perfection by the Arabs was the planispheric astrolabe. It was held in the vertical by a cord attached to a ring or lug at the top. The central disc had a raised rim on the front (the face) so that a plate inscribed with circles and curves centred on the North Pole could be placed in it. An appropriate plate was needed for each markedly different terrestrial latitude. On top of the plate was placed an open-work piece of metal, the rete, showing the ecliptic and pointers to a number of major stars, and on top of this a rotatable rule (see Fig. 10.5). On the back (dorsum) was an outer circle graduated in degrees; a second circle, divided into the signs of the zodiac and inside that into months and days of the month, had a movable alidade. Degrees and hours were shown on the rim of the front. A pin through the rule and the centres of the rete and the plate in front, through the centre of the main disc and the alidade on the back, was fastened by a wedge. Sighting holes were drilled through the projecting lugs on the alidade which when rotated to enable a star to be seen through the sighting holes also rotated the rule on the front. The first description of a planispheric astrolabe in English was given by Chaucer who also gave an account of many of its uses. For an extensive treatment of astrolabes over a long period of time consult Gunther (1976).

The planispheric astrolabe was a very versatile instrument. It could be used to establish the longitude and latitude of a celestial body on any given occasion. This use included the longitude of the Sun at given hours with given dates and at the equinoxes and solstices. It could also be used for estimating the heights of towers. It was both a sighting and a computing device.

Most of the sighting devices had inherent errors which militated against the establishment of accurate observational data. For example, a gnomon casts both an umbra and a penumbra (as a result of the Sun having a perceptible disc). Short distances between backsights and foresights allow substantial differences in observed direction. Ulugh Beg and Regiomontanus, both in the fifteenth century, Tycho Brahe in the late sixteenth century and Jai Singh in the seventeenth century built enormous quadrants and related instruments in order to achieve more accurate positional information. Tycho Brahe perhaps more than any of his predecessors recognized the need for frequent observations in a large range of situations. His data on Mars enabled Kepler to discover, after tedious and usually frustrated efforts, his three laws of planetary motion.

In 1900 a team of sponge divers discovered an ancient wreck off the coast of Antikythera, a Greek island. The statuary and pottery aboard was later dated to about 80 BC. Amongst the materials sent to the National Archaeological Museum, Athens, was an amorphous lump of metallic material which on drying out separated into several layers. It consisted of bronze sheets cut into rectangular plates, dials and gears. The whole complex was badly corroded and upon cleaning much of it was apparently destroyed. Derek de Solla Price (1957 and 1974) has made a painstaking study of the ancient instrument; his investigation seems to have been a very perceptive piece of detective work.

There are many indications, most incomplete, that this instrument was astronomical and that it may have been some sort of computer. The incomplete dials on front seem to have been divided into twelve segments, one labelled with the Greek name for *Libra* and another with part of the Greek name for *Virgo*; the remaining segments are divided into thirtieths – the Egyptian months each had 30 days. Another movable dial seems to have been changed every four years for the late entry (in terms of the Egyptian calendar of 365 days) of the Sun into the sign *Aries*. Inscriptions elsewhere suggest that letters on these plates indicate the dates of the risings of several stars, asterisms or constellations. Inscriptions on the back plate refer to the (near) equivalence of 235 synodic months and nineteen years, the Metonic cycle, and the (nearer) equivalence of 940 synodic months to 76 years (the Kallippic cycle). There also seems to be a reference on this plate to the 223 synodic month eclipse cycle.

Even more interesting are the numerous triangular toothed gears, some 32 remaining, usually in a poor state of preservation. The order of their placement is uncertain because some have been detected only by transverse radiography, but it seems that rotation of a contrate gear (a driver with teeth at right angles to its plane) would have generated a number of conjoint successive rotations in successive gear chains. Amongst these were (i) 254 sidereal months in nineteen solar years and (ii) through a much longer gear chain 235 synodic months in nineteen solar years. Other gearing, most of which ends in what seems to have been frustrated, may have had reference to other planetary periods.

Interestingly, some later Arabian astrolabes had triangular toothed gearing used for computational purposes. It is odd that no Hellenistic literary reference to such devices has been preserved. Possibly triangular toothed gears did not work well; the Antikythera device appears to have needed mending not long after its original construction.

CHAPTER ELEVEN

THE REAWAKENING OF THE WEST

As set out in Chapter 7, Europe went into intellectual decline from about the third century AD. This was more marked in the Western (Latin) empire than in the Eastern (Greek) empire. Most of the earlier intellectual contributions to mathematics, logic, astronomy, physics, metaphysics, biology and medicine had been made in Greek. The Romans, as reported above, at best summarized the major contributions or collated them in encyclopaedic compendia such as Pliny's *Natural History* (first century AD). In late Republican and early Imperial times the well-educated Roman citizen could and did read Greek. The Roman Church and its leaders became increasingly dependent on Latin, which was the language of the liturgy, of Jerome's Vulgate translation of the Bible (fourth century) and of ecclesiastical administration. As the Western and Eastern churches drifted apart, ending in the tenth-century schism, a knowledge of Greek became a rarity among allegedly learned western churchmen. By the ninth century, John Scotus Erigena, an Irish monk, was one of the few western scholars who had a sound command of Greek. There were, of course, few Greek texts for this small band of western scholars to work on.

By the eighth century a number of scholars began to take an increasing interest in what may be taken as temporal as distinct from doctrinal theological matters. One such was the English monk Bede (late seventh, early eighth century), who was under some influence from the long-isolated Irish church. He wrote a history of England, admittedly called a history of the church in England, and discussed several calendrical matters.

Later medieval Latin scholars began to take increasing notice of the Arabic scholars on their doorstep in Spain. At first they began translating into Latin Arabic translations of Greek manuscripts or perhaps often Arabic translations of Syriac translations of these Greek texts. Gerbert, a mathematician and later Pope Sylvester (late tenth century), Athelhard of Bath (early twelfth century) and Gerhard of Cremona (late twelfth century) were some of these translators. With the passage of time, as a result of a greater availability of Greek manuscripts brought by eastern Christians escaping Turkish persecution, and of an increasing knowledge of Greek also contributed by these refugees, more and more of the Latin translations were made from the Greek.

In these ways late medieval scholars in the Latin west became acquainted with Euclid, Ptolemy, Aristotle and a number of Greek contributors to mathematics, logic, astronomy, cosmology, physics, metaphysics and biology. At first they dipped their toes cautiously into these newly discovered waters. They were perhaps overawed by the apparent wisdom and authority of the ancients but they still saw the need to reconcile the Scriptures and these secular teachings. Thomas Aquinas, thirteenth century, a theologian and philosopher, became one of the most adept in the subtle art of reconciliation. He distinguished between beliefs held on faith and those held on reason. Further, there were the worrying inconsistencies amongst the ancient authorities. As al-Bitruji and other Arab astronomers had noted, there was an apparent incompatibility between Aristotle's physics and Ptolemy's circles upon circles. What was to be taken literally and what instrumentally (or as aids to analysis) were matters of doubt.

The late medieval Western scholars with an interest in astronomy had an increasing inclination to look into the horse's mouth, in this case the sky. Unfortunately they tended not to recognize that repeated observation with the best techniques available is necessary to reduce errors of observation and that the reported observations of ancient predecessors were subject to errors of observation or were sometimes 'adjusted' to fit a theoretical preconception.

By the thirteenth century scholars such as Grosseteste (c.

1168–1253) and Albertus Magnus (*c.* 1200–80) had a substantial mastery of Aristotelian philosophical and scientific writings and of Ptolemaic astronomy. Grosseteste was an important supporter of empirical, including experimental, methods in science. He recognized the difference between deductive proof as in geometry and the formulation on the basis of observations of hypotheses which could be tested by the observation of events they implied (see Crombie 1953). Grosseteste compared the estimates of the tropical year given by Hipparchos, by Ptolemy, by al-Battani and by Thabit ibn Qurra with such data as were available to him. He judged al-Battani's estimate to be the best. He considered Ptolemy's planetary theories in terms of deferents, epicycles and equants to be superior in respect of predictive power, but he admitted that al-Bitruji's theories were superior in terms of Aristotelian physics. This dilemma between an adequate predictive mathematical model and an acceptable physical statement had troubled Greek astronomers to some degree and to a greater degree many Arabian astronomers and many late Renaissance western astronomers.

Albertus Magnus was a diligent observer and insightful interpreter, for instance he suggested that the Milky Way was made up of faint stars and that the markings on the Moon were shadows of elevations (mountains) on its surface. He recognized the relevance of the Moon for the tides. Albertus like Grosseteste was suggesting empirically testable hypotheses concerning celestial events.

A near contemporary, Johannes de Sacrobosco or John of Holywood (died *c.* 1250) wrote an elementary astronomical text, *De sphaera*, which was used in the universities for some three centuries. It set out several spherical reference systems such as the celestial equator and the meridians crossing it at right angles, the ecliptic and latitudinal displacements from it. It described the zodiac and so on. As it proceeded with the largely qualitative information, it dealt finally with the five star-like planets. Sacrobosco seems to have known about Ptolemaic deferents, epicycles and equants for he says that each planet had its own values for each of these parameters, as we would call them. However, he does not specify what these terms mean nor does he give any of the values. In a way it illustrates the lack of detail and precision which must have marked university courses in astronomy in the thirteenth and immediately following centuries. This treatise remained in use until the sixteenth century.

Campanus of Novara (died 1296) wrote a technical treatise, known as *Theorica planetarum*, and a shorter popular *Tractatus de spera*. The latter is similar in plan to Sacrobosco's *De sphaera*. The former is quite technical, setting out Ptolemy's planetary models including the relevant quantities. It includes directions for the construction of an instrument which would later have been called an equatorium. From it, the positions of the seven planets at any given time could be read off with approximate accuracy. It was a series of discs which could be rotated. Campanus also wrote a shorter work on the quadrant; another short work on the astrolabe has been attributed to him. He also adapted the Toledan tables to Julian years, the Christian era and the meridian of Novara.

Richard of Wallingford (*c.* 1292–1336), who had studied and taught in Oxford, built an equatorium with which to calculate planetary positions. He was quite ingenious in the design and construction of astronomical instruments. One of his outstanding instrumental achievements was the design and construction of a mechanical astronomical clock at St Albans where he was abbot.

Nicole Oresme (*c.* 1320–82), who had studied with the great physicist Jean Buridan in Paris, metaphorically likened the heavens to a complex mechanical clock, the motions of the several bodies being sustained by impetus originally imparted by God. He drew attention to the relativity of the detection of celestial motion: as a boat sails away, a passenger sees the shore as receding. Though the celestial bodies seem to have an east-to-west diurnal motion, Oresme stated that this could be the result of a west-to-east diurnal rotation of the Earth. He opposed astrology, claiming that events on earth were the result of natural causes and not celestial influences.

Much of the early Western European interest in astronomy sprang from the obvious need to reform the ecclesiastical calendar, which was derived from the Jewish luni-solar calendar matched to either the Roman Julian calendar or the Alexandrian Egyptian calendar. The two main problems in respect of Easter, on the date of which the dates of all the other variable feasts depended, were, as we have seen, (i) the length of the tropical year, with which was associated the rate of the precession of the equinoxes (an issue bedevilled by Ptolemy's error and the later erroneous doctrine of trepidation), and (ii) the prediction of the dates of new Moon, which is complicated by the slight variation in the apparent motion of the Sun and the marked variation in both the sidereal period of the Moon and its anomalistic period. Concern with the ephemerides of the star-like planets was relevant to astrology, which was in vogue in the late medieval West.

In the fifteenth century we find several important figures working on the bases laid down in the thirteenth. Amongst these were Nicholas of Cusa (c. 1401–64), who contributed more to philosophy and mathematics than directly to astronomy, Georg Peuerbach (1421–61) and Johannes Müller (1436–76):

Nicholas of Cusa was a speculator about and critic of various astronomical, or better, cosmological, questions. He went on to consider that the Earth might be in motion. He had no evidence that it was but he was willing to question the assumption that it could not be. Peuerbach did most of his astronomical work in collaboration with his disciple Müller, usually called Regiomontanus after his birthplace Königsberg. They observed what we now know to be Halley's comet in June 1456. They observed a lunar eclipse on 3 September 1457 and found the time of mid-eclipse to be eight minutes earlier than predicted from the Alphonsine tables. They found the obliquity of the ecliptic to be 23°28′. Peuerbach produced a set of eclipse tables and began an abridgement of the Latin version of the Arabic *Almagest*. He is said to have known the version by heart but doubted its faithfulness to the original Greek, a language which he set out to learn. He was encouraged here by Bessarion, a Greek Orthodox dignitary who came to Rome in an attempt to unite the eastern and western churches. Bessarion was successful in Rome but the Orthodox Church would not ratify the agreement. Bessarion joined the Catholic Church, was appointed a cardinal and was used by the reigning Pope on a number of diplomatic missions, on one of which (to Vienna) he met Peuerbach and Regiomontanus. It was Regiomontanus rather than Peuerbach who mastered Greek and Bessarion supplied him with numerous Greek manuscripts which he had brought from Constantinople, already under final pressure from the Turks.

Peuerbach wrote an elementary but careful textbook, *Theorica novae planetarum*, published in 1454; Regiomontanus produced an enlarged version in 1474. Peuerbach had not finished his epitome of *Almagest* at the time of his early death. Regiomontanus completed the task making use of Greek manuscripts.

Regiomontanus improved several sighting devices such as the Jacob's staff, the triquetrum and the quadrant, mainly by increasing their size. These were used by Regiomontanus and his young collaborator Bernhard Walther (1430–1504) to establish better data on solar altitudes and planetary and stellar positions. In roughly a quarter of a century Walther, at first in collaboration with Regiomontanus, had established 746 solar altitudes and 615 planetary and stellar positions. A century or so later Tycho Brahe and Kepler made use of these Regiomontanus-Walther data.

Regiomontanus was not only an instrumentalist and methodical observer. He saw the need for more precise mathematical analysis of data and to this end constructed a new set of trigonometrical tables. He may even have speculated that the Earth was in motion. He is alleged to have written in a letter, now lost, that the motion of the stars must vary a little because of the motion of the Earth; if this report is correct he also recognized that the revolution of the Earth would result in a stellar parallax, which until the 1830s had not been observed.

Celio Calcagnini (1479–1541), who spent a period of time during 1518 in Cracow, where Copernicus had earlier studied, supposed the Earth to be rotating west to east in about 24 hours rather than the heavens to be rotating east to west; he seems not, however, to have suggested that the Earth revolved around the Sun rather than vice versa. Copernicus had announced both propositions *circa* 1507.

Girolamo Fracastoro (1478–1553) and Copernicus were at Padua together from 1501. Fracastoro became dissatisfied with the Ptolemaic theories, not because they were 'geocentric' but because the excentrics, the epicycles and the equants could not be reconciled with Aristotelian physics. Consequently he attempted in 1538 to reinstate what was basically a Eudoxan theory of homocentric spheres, but with their axes at right angles to one another. In general, a set of five spheres was used for both the fixed stars and each planet. Trepidation of the equinoxes was accepted and explained by the interaction of the second and third spheres governing the fixed stars. All told, Fracastoro had 77 spheres, clearly a retrogression in terms of economy of assumption.

Thus in the late fifteenth and early sixteenth centuries we find in the West (i) a clearer understanding of the Eudoxan-Aristotelian and of the Ptolemaic planetary theories, (ii) a puzzlement over whether planetary theories should describe nature or whether they should be merely an aid to prediction, and (iii) a recognition of the need for better data and of the relevance to this objective of more numerous observations made with improved observational aids. All these are to be found in Arabian astronomy, in which no revolution occurred. The Copernican revolution, which in many respects was quite conservative, was scarcely prompted by any new facts. It was mainly prompted by dissatisfaction with the Ptolemaic planetary theories, as originally stated and as elaborated and patched up. The central motivation, however, seems to have been Copernicus's concern with Ptolemy's breaches of the 'Platonic' requirement of reduction of anomalous planetary motions to component uniform circular motions. He considered that the major Ptolemaic breaches were the equants and the crank-mechanisms. Copernicus failed to recognize some of the advantages of his own revolutionary theory, partly because he tied it down with so many conservative restraints such as component uniform circular motions.

In relation to Copernicus it is important to note that when he was a student at Cracow in the late fifteenth century there was a strong school of astronomy there and that he enrolled in several astronomical and astrological courses. In addition he seems to have had extensive out-of-class contacts with Wojciech Brudzewski, a noted mathematician and critic of Ptolemaic astronomy. Later in Bologna he collaborated with Domenico Maria Novarro (1454–1504). A joint observation by them of an occultation by the Moon of Aldebaran led them to question Ptolemy's geometric theory of the Moon.

CONSTELLATION AND STAR NAMES
FROM ANCIENT TO MODERN TIMES

It was claimed in Chapter 2 that most, if not all, of the names of the zodiacal constellations and of the associated signs of the zodiac originated in Mesopotamia and were translated into or otherwise adapted in Greek, Latin, Sanskrit, Southern Indian languages, Arabic and Iranian (see Table 12.1). Though the Chinese used animal names for the signs of the zodiac, they adopted completely independent names.

By contrast the planetary names were markedly different from culture to culture. It is difficult to say what the Sumerian and Akkadian names meant, though *Gud.ud* in Sumerian may have meant 'the calf of the Sun' as Mercury never strays far from the Sun, and *Marduk* in Akkadian is the name of the chief god in the Babylonian hierarchy; *Nin.dar.anna* (sometimes rendered *Nin.-si.anna*) in Sumerian probably meant 'the multi-coloured or bright mistress of the heavens', and *Ishtar* in Akkadian was the goddess of love. The Greek and Latin names of the five star-like planets are in parallel. *Hermes, Mercurius* is a messenger god; *Aphrodite, Venus* is a goddess of love; *Ares, Mars* is a god of war; *Zeus, Iuppiter* is the chief god in the divine hierarchy, and *Khronos, Saturn* is his

Table 12.1 The names of the zodiacal constellations in several languages and their meanings (in some cases there are two names)

Sumerian	Akkadian	Greek	Latin	Sanskrit	Persian	Arabic
Lu-hun-ga (farm-worker)	Agru (farm-worker) Immeru (sheep)	Krios (ram)	Aries (ram)	Mesha (ram)	Barra (lamb)	Hamal (ram)
Gu.ud an.na (bull-of-heaven)	Alap-same (bull-of-heaven)	Tauros (bull)	Taurus (bull)	Vrishaba (bull)	Gav (bull)	Thaur (bull)
Mas.tab.ba gal.gal (great twins)	Masu (great twins)	Didymoi (twins)	Gemini (twins)	Mithuna (twins)	Du-payker (two-figured)	Jauza (twins)
Al.lul (crab)	Allutu (crab)	Karkinos (crab)	Cancer (crab)	Karka (crab)	Khar-chang (crab)	Saraton (crab)
Ur.gu.la (?great dog)	Ka.lab.me.e (lion)	Leon (lion)	Leo (lion)	Sinha (lion)	Shir (lion)	Asad (lion)
Ab.sin (furrow)	Absinnu (furrow)	Parthenos (virgin)	Virgo (virgin)	Kanja (virgin)	Khusha (ear of wheat)	Sumbola (wheat sheaf) Adhra (virgin)

Zib.ba.anna (heavenly weighing scales)	Zibanitu (weighing scales)	Chelai (claws of the scorpion) Zugon (beam)	Libra (weighing scales)	Tula (beam)	Tarazu (scales)	Zubawa (claws) Mizam (beam)
Gir.u.tab (stinger)	Agrabu (scorpion)	Skorpios (scorpion)	Scorpio (scorpion)	Vriscika (scorpion)	Gazh-dum (scorpion)	Aqrab (scorpion)
Us. (soldier) Pa.bil.sag (?archer)	Nedu (soldier)	Taxotes (archer?)	Sagittarius (archer)	Dhanus (bow)	Kaman (bow)	Qaus (bow)
Suhur.mas (fish-goat)	Suhu-massa (fish-goat)	Aigokeros (goat-horned)	Capricornus (goat-horned)	Makara (sea monster)	Buz-i-kuhi (wild goat)	Jady (goat)
Gu.la (?great man)	Qu-hasbu (?streams)	Hydrochoos (water-pourer)	Aquarius (water-carrier)	Kumba (urn)	Dul-i-asiyah (water mill hopper)	Dalw (bucket)
Zi.me (tails) Du.nu.nu (fishcord)	Zibbati (tails) Rikis nu.mi (fishcord)	Ichthyes (fish)	Pisces (fish)	Mina (fish)	Mahi (fish)	Hut (fish)

Table 12.2   The names of the seven planets in several languages

English	Sumerian	Akkadian	Greek	Latin	Sanskrit	Arabic
Sun	Ud	Shamas	Helios	Sol	Sura	Shams
Moon	En.zu	Sin	Selena	Luna	Chandra	Qamar
Mercury	Gud.ud	Gut.tu	Hermes	Mercurius	Buddha	Zi'bag
Venus	Dilbad Nin.dar.anna	Na.baat Ishtar	Aphrodite	Venus	Sukra	Zuharah
Mars	Sa	Salbatanu	Ares	Mars	Angaraka	Mirrikh
Jupiter	Sag.me.gar	Nibiru Marduk	Zeus	Iuppiter	Brihaspati	Mushtari
Saturn	Sag.us	Kalmanu	Khronos	Saturnus	Sani	Zuhal

displaced father. The Sanskrit names are quite different: *Buddha* is the wise one, *Sukra* is the white one, *Angaraka* is the glowing ember (a reference to Mars's ruddy colour), *Brihaspati* is lord of prayer and *Sani* is the tardy one (a reference to Saturn's long sidereal period of almost 30 years). The Arabic names are probably derived from earlier times before there was any contact with Hellenistic astronomy but I do not know what meanings, if any, they had. These variant names are set out in Table 12.2.

In addition to their names for the zodiacal constellations and for the seven planets, the Mesopotamians had a name for the Milky Way (*Du.ran* in Sumerian and *Rikis.same* in Akkadian). The names of 46 non-zodiacal constellations and names of 341 individual stars or tight asterisms such as the Pleiades have been preserved, and are collated in Gössmann, *Planetarium Babylonicum*.

The Pleiades at first glance look like a rather fuzzy star but closer inspection by a keen eye on a good night reveals six or seven stars. The Greeks named them as the seven daughters of Atlas. Galileo's primitive telescope revealed many more in the cluster; the brightest called Alcyone by the Greeks is 3rd magnitude, five are 4th magnitude, three are 5th magnitude (but only one is far enough away from brighter members of the cluster to be detectable by a keen eye) and sixteen are 6th magnitude (the minimum brightness detectable by the naked eye, but in this case they are too close to brighter members to be distinguished). The Sumerians called this asterism *Mulmul*, the star of stars, and placed it on the tip of the northwestern crumpled horn of the *Bull-of-Heaven*. The Babylonians called it *Zappu*, the bristle or tuft of hair in the *Bull's* mane.

In Mesopotamia, the constellation called by the Greeks *Orion*, the hunter with club in his upraised right arm and a lion-skin draped from his left arm, was seen in much the same configuration as *Sitaddalu*, the shepherd. The lion-skin shield of *Orion* constituted the southern crumpled horn of the *Bull-of-Heaven*, and perhaps the club in the right hand may have been seen as the top of a crook (see Fig. 12.1).

Over the head of *Lu-hun-ga*, the hired farm-worker, later *Immeru*, the sheep (our *Aries*), was a triangle of stars called in Mesopotamia *Apin*, the plough; the Greeks called this simple constellation *Deltotan*, shaped like the letter *delta*, or *Trigonen*, the triangle. Our name for it is *Triangulum*.

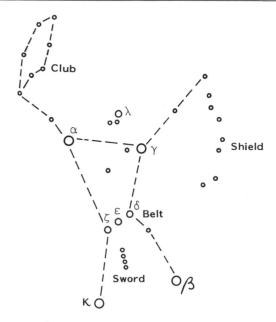

Fig. 12.1   *The central pattern or* imago *of the constellation* Orion. *The star* alpha *is Betelgeuze,* beta *is Rigel and* gamma *is Bellatrix. The stars deemed to constitute* Orion's *club, belt, sword and shield are indicated.*

In each of the three examples, the *Bull-of-Heaven,* the *Shepherd* and the *Plough,* there was a degree of congruence between the Mesopotamian and Greek conceptions of the constellation and in one case the name applied. In the case of *Canis Major,* the Mesopotamian conception and naming was quite different. The Mesopotamians saw this set of stars as two interlocked constellations: *Ban* the bow and *Kak-ban* the arrow (the latter with *Kak.si.di,* our Sirius, as its tip). While most Greek constellations resembled the Mesopotamian ones in conception and sometimes in name, many were quite different in one or both respects.

The referents of most of the 241 preserved Mesopotamian names for individual stars or asterisms can be identified, though in some cases there is a degree of uncertainty. Our Aldebaran was called *Ne.gi.ne.gar,* our Antares was *Lisi-gun,* our Canopus (*alpha Carinae*) was *Nunki,* our Hyades was *Sis.da,* and so on. It is difficult to establish the meanings of most of these Mesopotamian names.

The succession over the centuries of names of *alpha Leonis*, the brightest star in the zodiacal constellation *Leo*, is worthy of mention. In Mesopotamia, according to Allen who is not always to be trusted, it was called *Sharru*, the king; in Greece it was *Basiliskos*, the little king; in Rome, *Rex*, the king, or *Regulus*, the little king (Copernicus also called it *Regulus*). The Greeks also called it *Kardia leonitos*, the heart of the lion, because of its position in the constellation. This became *Cor leonis* in Latin and *Qalb al-Asad* (the heart of the lion) in Arabic. In late medieval Europe this Arabic name was corrupted into *Kalbelaset*, *Calb-alezet* and other variants in the astrolabe star lists.

Kak.si.di, the tip of the Mesopotamian arrow, was *Spdt* (perhaps *Sopdet* when we supply the missing vowels) in Egypt. Its heliacal rising was used to mark the opening of the year of the seasons and to enable a rough prediction of the rise of the Nile. The heliacal rising of *Spdt* usually differed by no more than a few weeks from the rise of the Nile. The intervals between successive heliacal risings of this star as observed from Memphis averaged about 365.2507 days, a period much closer to the tropical year of say 365.2424 days *circa* 1500 BC than the Egyptian civil calendar year of 365 days. The first day of the civil calendar year worked its way forward through the year of the seasons in about 1506 calendar years, whereas the rise of the Nile would have varied by a few weeks according to the date of onset and the volume of rain in Ethiopia feeding the Blue Nile.

Though the Greeks called the Egyptian *Spdt Seirios*, the scorcher, they transliterated the Egyptian name into *Sothis* and spoke of the slow working of the beginning of the Egyptian calendar year forward through the year of the seasons as the Sothic cycle; because they used a slightly erroneous value for the tropical year they calculated the cycle to be 1460 Egyptian calendar years whereas it is almost 1506 calendar years. Sirius over a long period made its heliacal rising in lower Egypt in mid- to late summer, hence it was for many centuries a good marker of the beginning of the year of the seasons. The Romans sometimes called Sirius *Canicula*, the little bitch, though that name was later sometimes used for Procyon, *alpha Canis Minoris*.

In Ptolemy's star catalogue there were 21 northern and fifteen southern constellations in addition to the twelve zodiacal constellations, in total ten less than the 58 Mesopotamian constellations reported by Gössmann. These 58 were not necessarily recognized at the same point in time; I cannot make a decision on this, for though Gössmann's reported information is extensive, in detail it is usually cryptic. Most of Ptolemy's constellations, as well as having a central pattern or *imago*, had a number of neighbouring stars outside that *imago*. Thus in the *Bull* there were 33 stars in the *imago* and eleven outside it; in the *Lion* there were 27 stars in the *imago* and five outside it; and so on. Ptolemy's 21 northern constellations have been increased to 29. *Serpens* has been divided into *Serpens Caput* and *Serpens Cauda* and seven of the sets of stars on the fringe of Ptolemy's *imagines* have been given separate status as constellations; amongst these are *Camelopardus*, *Canes Venatici* and *Coma Berenices*. Ptolemy seems not to have included stars beyond 53° south declination in his time, hence as more stars between 53° south and the South Pole were discovered, additional constellations have been created. Further, some Ptolemaic southern constellations have been subdivided. Ptolemy's *Argo*, the ship, has been divided into four – *Carina*, the keel, *Pyxis*, the compass, *Puppis*, the stern, and *Vela* the sail. Also *Crux* has been separated from *Centaurus*. Ptolemy's fifteen southern constellations have been extended to 44.

In Ptolemy's star catalogue, only seven individual stars were listed with proper names additional to their locations within the constellations. Arcturus, the bear-watcher, Antares, the rival of Mars, Procyon, the preceder of the dog, and Canopus, named after a city in the Nile delta, are Greek names; Capella, the little she-goat, Spica, the ear of wheat, and Regulus, the little king, are Latin. More usually Ptolemy gave his stars only a location in a constellation and an ecliptic longitude and latitude. Thus the bright red star of the Hyades said to be in the southern eye of the *Bull*, $10\frac{5}{6}°$ of the *Bull*, $5\frac{1}{6}°$ south, is our Aldebaran; the brightest and red star in the face of the *Dog*, $17\frac{2}{3}°$ of the *Twins*, $39\frac{1}{6}°$ south, is our Sirius; the bright red star in the right shoulder of *Orion*, 27° of the *Bull*, 17° south, is our Betelgeuze; and so on.

In addition to the seven Greek and Latin names listed in Ptolemy's star catalogue, some others from these sources came into general Western usage. Thus Sirius, Castor, Pollux and Alcyone (the brightest star in the Pleiades) are Greek; Bellatrix, the female warrior (in the left shoulder of *Orion*), Mira (a variable star in *Cetus* and hence regarded as a wonder), Vindemiatrix, the woman grape-picker (in *Virgo*), and Polaris (a modern labelling of the present pole-star) are Latin. At least one Mesopotamian name is in use, Nunki, though we have transferred it from Canopus to *sigma Sagittarii*.

Most proper names in recent use are of Arabic origin. Kunitzsch (1983) claims three types of such origin. First, there are names derived from early Arabic usage prior to Arabian contact with Hellenistic astronomy. Amongst these are Aldebaran from the Arabic *Al-Dabaran*, the follower (of the Pleiades, which rose earlier), Adhara from *Al-Adhara*, the virgins, and Alphard from *Al-Fard al-Shuja*, the solitary one in the water-snake, *Hydra*. Betelgeuze is another example, though manifesting serious misunderstanding and resultant corruption. We regard it as marking the shoulder of *Orion*. The early Arabic name for this star was *Yad al-Gawza*, the hand of *Gawza* (our *Orion*). There is some disagreement about the meaning of *Gawza*; it may have been 'giant' but it is more likely to be 'central one'. The Arabic *yad* was correctly read by European scholars in the thirteenth century. Later scholars misread 'y' as 'b', producing *Bad al-Gawza*. *Bad* is not, according to Kunitzsch, an Arabic word, so some later scholars considered it was a corruption of *Ibt*, armpit.

Second, and more numerous, are star names derived from Arabic translations of Ptolemy's description of the star's position in its constellation. Thus Ptolemy described a star as being in the mouth of the (southern) fish; the 'mouth of the fish' in Arabic is *Fam al-Hut*, yielding our Fomalhaut. As we have seen, *Qalb al-Asad*, the heart of the lion, was corrupted in late medieval Europe to *Kalbelaset, Calbalezet* and other forms; the current names are Regulus and Cor Leonis. The bright star in the head of the Gorgon became *Ras al-Ghul*, the head of the demon, which became Algol. The star at the end of the river Eridanus became *Al-Ahir al-Nahr* and was corrupted to Achenar. The brightest star in the head of the Ram, *Ras al-Hamal*, the head of the sheep, was reduced to Hamal, and *Rijl Gawza al-Yura*, the right foot of *Gawza*, was reduced to Rigel.

In the same vein the Arabian astronomers followed Ptolemy in placing the second brightest star in *Orion* in the shoulder and consequently called it *Mankib al-Gawza*, from which Western astronomers derived Mancamalganze and Malgeuze, but these ultimately gave way to Betelgeuze with variant spellings such as

Badalgeuze, Beitalgeuze, and 'genze' sometimes being substituted for 'geuze'.

The third set of names derived from the Arabic were bestowals, often ill-based, by early modern Western astronomers even though they had never been used by Arabian astronomers. Most of these names have disappeared. Thuban, *alpha Draconis*, is an exception.

Because astronomers in the last century have adopted a more systematic way of identifying stars, this has led to a diminution of proper names. The system will be described below.

Allen (1899) is a mine of information on constellation and star names in numerous languages. It is obviously deficient in respect of Mesopotamian names but can be supplemented by Gössmann. Though Allen has a great deal of material on Arabic names and a substantial amount on Persian names, Kunitzsch warns that this has to be treated with caution. For instance Allen claims that Bad al-Gawza and *Yad al-Gawza* are corruptions of *Ibt al-Gawza*. Again, the Arabians called *alpha*, *beta* and *gamma Aquilae* by a word meaning 'the balance'. The Persian translation of the Arabic was *shalim-i tarazu*, which was misread as the preying falcon and was so reported by Allen. The first part of the Persian name, according to Kunitzsch, was Arabicized to *Alshain*, and the second to *Tarazed*, and bestowed as proper names on *beta* and *gamma Aquilae* respectively.

From the eighteenth century onwards Western astronomers began putting the constellations within firm boundaries defined on some co-ordinate system such as meridians of Right Ascension measured along the celestial equator from the spring equinox (or First Point in Aries) and as parallels of equatorial declination. They tried to enclose within these boundaries not only the stars in the traditional constellation figures but also what had been called neighbouring stars outside the figures. They attempted to designate the stars within such a constellation in descending order of brightness by means of Greek letters, numbers and Roman letters but to begin with there was no generally agreed set of constellation boundaries and no agreed set of star designations within a constellation. By international agreement in the nineteenth century the constellation boundaries were agreed upon, as was the ordinal designation of stars in each constellation. The constellations, often with jagged edges (see Norton's *Star Atlas*), preserve their early modern proper names such as *Ursa Major, Taurus, Virgo* and *Centaurus*. The stars within each constellation are given, in approximately descending order of magnitude, Greek letter designations, followed by numerical designations. These less personal designations have tended to replace the proper names in earlier use. Thus a modern astronomer is more likely to refer to *alpha Virginis* than to Spica, to *gamma Orionis* than to Bellatrix, to *beta Leonis* than to Denebola and so on.

Indian astronomers from perhaps the first half of the first millennium AD used names derived from the Greeks for the zodiacal constellations, either Sanskrit translations (or equivalents) of the Greek or Dravidian corruptions of the Greek names; thus we have in northern India *Mesha*, the ram, the Sanskrit equivalent of the Greek *Krios*, whereas in southern India we have *Kreya*, an obvious corruption of the Greek *Krios*. Some other names of Greek origin were used for other constellations but in addition many indigenous conceptions and names were used. Though the Indians placed great stress on the twelve zodiacal constellations of the Mesopotamians and the Greeks, they placed as great a stress on the 27 or 28 *Nakshatra* stars, the successive brides of the Moon, unfortunately like the Mesopotamian reference stars and the Chinese 'palaces' of the Moon, markedly unevenly spaced. The Moon progresses by only slightly variable steps amongst the fixed stars: the average is $13°10'37''$ per day with a minimum of about $11°$ and a maximum of about $15°22'$. The first *Nakshatra* was *Krittikah*, the woven, our Pleiades; the second was *Rohini*, the red deer, our Hyades; the fourth was *Ardra*, the moist one, our Betelgeuze; the fifth was *Punarvasu*, our pair Castor and Pollux, and so on. These names are still in popular use and are the bases of the Indian lunar month names. The Western system of constellation limits and Greek letters and Arabic number designations of individual stars has been adopted in recent times for scientific purposes. Renou and Filliozat (1953) is a useful but not exhaustive source on Indian constellation and star names.

As we shall see, Chinese conceptions of the constellations and the names bestowed on them and on their constituent stars are in the main different from those used in the Mesopotamian, Greek, Indian and Arabian tradition.

COPERNICUS

Mikołaj Kopernik (AD 1473–1543), better known to us by the Latinized form of his name, Nicolaus Copernicus, is often regarded as the first great modern astronomer. In replacing the traditional geocentric (more exactly geofocal) treatment of the universe inherited from the Greeks, he certainly opened the door for modern heliofocal planetary theory as developed by Kepler and Newton. His design of the door and its controlling mechanisms, however, was so much in the ancient tradition exemplified by Hipparchos, Ptolemy and the Arabs that it may reasonably be said, as Kopal (1973) did, that Copernicus was the last great ancient astronomer. Copernicus saw a way in this ancient tradition to work out in reasonable detail the hypothesis which Aristarchos and Seleukos, almost two millenniums before, seemed able to state only in general terms. They had supposed that the Earth was in daily rotation around its axis and in annual revolution around the Sun, and that the other planets, except the Moon, revolved around the Sun and not around the Earth. If Aristarchos or Seleukos worked out any mathematical details, these details have not been preserved. Copernicus succeeded in just this task, at least to a reasonable approximation to the observed situation.

His great contribution was to assume an alleged orbit of revolution of the Earth around the Sun in place of Ptolemy's hypothesized orbit of the Sun around the Earth. This enabled him to bypass Ptolemy's main reason for assuming his deferents of Mercury and Venus and his epicycles of Mars, Jupiter and Saturn because the Earth's orbit served the same geometrical functions as

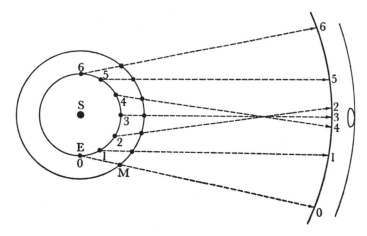

Fig. 13.1 *Copernicus's explanation of the retrograde phase of a planet, illustrated by Mars. S is the Sun and the segment of a circle to the right is part of the celestial sphere. As the Earth (E) catches up with and overtakes Mars (M), the latter appears to go forward, then backwards (from points 2 to 4) and then forward again.*

Ptolemy's deferents of the first two and his epicycles of the three last.

Copernicus had further to assume that the orbits of Mercury and Venus around the Sun lay inside the orbit of the Earth and that the orbits of Mars, Jupiter and Saturn lay outside that of the Earth. As a result the first two came to be called inferior planets and the three last superior planets.

Several important consequences flow from these assumptions of Copernicus. First, Copernicus did not need Ptolemy's epicycles in order to account for the long phases of apparent direct motion interspersed with short phases of apparent retrograde motion. A more distant planet would be overtaken periodically by a less distant planet so from the Earth a superior planet would periodically seem to retrograde (see Fig. 13.1) and inferior planets would never be seen far from the Sun (see Fig. 13.2).

Second, they enabled Copernicus to eliminate some unexplained coincidences in the Ptolemaic planetary theories. In Ptolemy's theories, the radius vectors of the deferents of Mercury

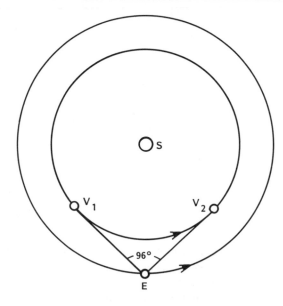

Fig. 13.2   *In terms of Copernicus's heliofocal theory inferior
planets such as Venus never stray far ahead of or behind the
Sun.*

and Venus lay on the radius vector of the Sun and the radius
vectors of the epicycles of the other three planets were parallel to
that of the Sun (see Fig. 6.4). In substituting the orbit of the Earth
for Ptolemy's orbit of the Sun, Copernicus removed these
coincidences.

Third, if stated in modern terms of reckoning the motion of a
planet around the circumference of a Ptolemaic epicycle relative
to the motion of the epicycle's centre around the deferent, the
periods of the deferents of Mercury and Venus and the periods of
the epicycles of the other three, are each one year (or the period
of the Earth's orbit), whereas the periods of the epicycles of the
first two and the periods of the deferents of the other three were
what we call the periods of revolution of these planets around the
Sun. Thus, Copernicus substituted the orbit of the Earth both for
Ptolemy's deferents of the two inferior planets and for his
epicycles of the three superior planets.

Table 13.1 Copernicus's values for the mean distances of the planets from the Sun using the Earth's mean distance as the unit value

Planet	Copernicus	Modern value	Ptolemy's ratios
Mercury	0.3763	0.3871	0.3708
Venus	0.7193	0.7233	0.7194
Earth	1.0000	1.0000	—
Mars	1.5198	1.5237	1.5191
Jupiter	5.2192	5.2028	5.2165
Saturn	9.1743	9.5389	9.2336

In the column headed Ptolemy's ratios, for Mercury and Venus the ratio is that of the epicycle to the deferent whereas for the other three planets it is that of the deferent to the epicycle.

Fourth, and perhaps most important, Ptolemy had to select arbitrary ratios of the radii of his deferents and epicycles in order to make his geometrical analyses work. He was free, however, to do so independently in respect of each planet. Copernicus in tying the orbit of the Earth into all his planetary analyses, except for that of the Moon, introduced a common factor and produced a unified theory. Copernicus like Ptolemy had to assume arbitrary ratios of the radii of the Earth and the other planets. His ratios were essentially the same as Ptolemy's (see Table 13.1). There was no way of falsifying Ptolemy's ratios except to show that they did not produce sufficiently good predictive results. In that case they could be adjusted in order to enable more accurate predicted positional values. As soon as it was possible to establish even approximately the relative distances of the planets from the Sun, it became possible to test (confirm or falsify, at least in general terms) Copernicus's assumed values.

Copernicus's theory of the Moon remained independent of his unified theory of the six planets (excluding the Sun but including the Earth).

In about 1514 Copernicus circulated to a few friends a manuscript known as *Commentariolus* (see Rosen's translation, 1959) which outlined the main features of his emerging planetary theory. At that stage he thought he could simplify the Ptolemaic

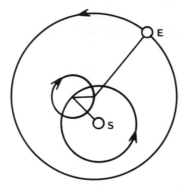

Fig. 13.3   *Copernicus's analysis of the orbit of the Earth involved a deferent and two epicycles.*

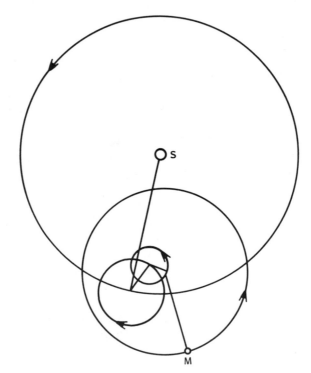

Fig. 13.4   *Copernicus's analysis of the orbit of Mars involved a deferent and three epicycles. Similar analyses with different radii and periods were used for Venus, Jupiter and Saturn.*

theories as well as remove some objectionable features such as excentrics and equants which he regarded as contrary to the supposed Platonic requirement that anomalous apparent planetary motion be reduced to uniform circular component motion. Copernicus also objected on the same ground to the crank-mechanisms used by Ptolemy in his theories of Mercury and of the Moon.

Copernicus worked out these basic ideas in greater detail in his *De revolutionibus* (1543). He could not dispense entirely with excentrics but he did produce constructions avoiding both equants and the crank-mechanisms. He had, however, to resort to deferent-epicycle couples and in most cases to additional epicycles on the primary epicycle. Thus he ended up with more circles on circles, not less than used by Ptolemy, despite his substitution of the Earth's orbit for five of the Ptolemaic circles – the deferents of Mercury and Venus and the epicycles of Mars, Jupiter and Saturn.

Whereas Ptolemy assumed that the Sun revolved around the Earth on an excentric circle (to account for the year of unequal seasons) and that the celestial sphere revolved around the Earth in a little less than a day, Copernicus had the Earth revolve around a secondary epicycle on a primary epicycle on a deferent around the Sun (see Fig. 13.3) – three circles to Ptolemy's one – and rotate around its polar axis which shifted in conical fashion, thus adding a fourth circle. Like Ptolemy, Copernicus accounted for precession by means of an additional circle.

For Venus, Mars, Jupiter and Saturn, Copernicus had three epicycles on the deferent (an increase of two circles for each, over Ptolemy's analysis) (see Fig. 13.4). These analyses including the ratios of the radii of the deferent are remarkably like those of al-Shatir about a century before Copernicus. There is no evidence that al-Shatir's analyses were available in a language known to Copernicus; the similarities strain any belief in either coincidence or independent invention (see Table 13.2). The Copernican theory for Mercury consisted like al-Shatir's of a deferent of three epicycles on the third of which was an al-Tusi couple (see Fig. 9.1). It has been suggested that Oresme may have understood the 'Tusi' couple but the suggestion is only tentative (see Kren 1971). It still has to be shown how Oresme could have become acquainted with al-Tusi's proposals.

Table 13.2 The ratios of the radii of the epicycles to the radius of the deferent according to al-Shatir (S) and Copernicus (C)

Planet		Deferent	Epicycles		
			First	Second	Third
Venus	S	60	1.683	0.433	43.550
	C	60	1.867	0.617	43.150
Mars	S	60	9.000	3.000	39.500
	C	60	8.767	3.000	39.483
Jupiter	S	60	4.125	1.375	11.500
	C	60	4.122	1.374	11.500
Saturn	S	60	5.125	1.708	6.500
	C	60	5.124	1.710	6.533

The relative ratios hold within the analysis for a given planet and not between planets. The deferent of each planet is assigned a value of 60, no matter what the relative distance of a planet from the Sun.

Copernicus's theory of the Moon involved a deferent centred on the Earth and two epicycles. The centre of the first epicycle moved west to east in a mean sidereal month. The centre of the second epicycle moved east to west in a mean anomalistic month. The Moon borne on the circumference of the second epicycle moved west to east in half a synodic month (see Fig. 13.5). This analysis eliminated the unfortunate implication of Ptolemy's analysis that the Moon varied by about a factor of two from apogee to perigee in its distance from the Earth and as a result varied comparably in its apparent size.

In all the above analyses Copernicus was trying to account for apparent motion in longitude. As bases for prediction of longitude they were little better than Ptolemy's. As we saw in Chapter 6, Ptolemy had great difficulty in accounting for variations in planetary latitude. He tried to do so by assuming tilts in the deferents relative to the ecliptic but his geocentric theories required him to have the planes of these orbits pass through the

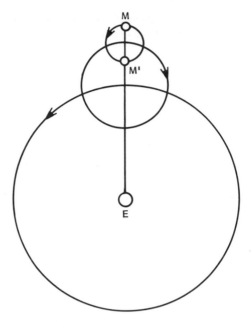

Fig. 13.5   *Copernicus's theory of the orbit of the Moon.*

Earth (when the planes of the planetary orbits in fact pass through the Sun). Had Copernicus had a truly heliocentric theory he could have been closer to the mark. He handicapped himself more than he needed to by passing the planes of the primary planetary orbits through the centre of the Earth's deferent which we have seen was offset from the Sun.

It is important to recognize that Copernicus's *De revolutionibus* is organized in respect of topics in parallel with Ptolemy's *Syntaxis mathematike*. It begins by rejecting Ptolemy's arguments that the Earth is stationary and at the centre of the universe. His arguments are not much, if any, better than Ptolemy's in support of the contrary propositions. Both were handicapped by their failure to understand inertia. Perhaps Copernicus was a little more culpable here in that Jean Buridan had already shed a little, if only glimmering, light on the matter through his doctrine of impetus. Greater illumination had to await what Newton could build on Galileo's conception of inertia. Copernicus went on to argue for the three alleged basic motions of the Earth: revolution around the Sun, rotation around its own axis, and the rotation of the polar axis relative to the fixed stars.

He next treated a variety of geometrical matters, including the major astronomical circles and the precession of the equinoxes. In the midst of these preliminaries, he like Ptolemy set out a star catalogue, almost entirely an updated version of Ptolemy's. He then proceeded to his analyses of the Sun's and the Moon's apparent motion and followed with his analyses of the motions of the five star-like planets in longitude and finally in latitude.

Copernicus in many instances was misled by errors made by his predecessors, especially Ptolemy. Taking the longitudes of certain stars reported by various observers, Copernicus concluded that precession had occurred at the rate of 1° in 79 years between *circa* 290 BC (Timochares) and *circa* 130 BC (Hipparchos), 1° in 100 years between *circa* 130 BC and *circa* AD 137 (Ptolemy), 1° in 85 years between *circa* AD 137 and *circa* AD 887 (al-Battani), and 1° in 71 years between *circa* AD 887 and *circa* AD 1520 (Copernicus). Had he ignored Ptolemy's alleged observations he would have arrived at a value of 1° in about 72 years. As it was, Copernicus wrongly concluded that the rate of precession fluctuated. Modern evidence indicates that the rate of precession is constant at 1° in a little more than 71 years.

As we have learned, Eratosthenes in the third century BC assessed the obliquity of the ecliptic as between 23°51′ and 24° (say 23°55′30″, an overestimate). Ptolemy reported that his observations confirmed the lower of Eratosthenes's values (though in his epoch some five centuries later, that would have been a still greater overestimate). Several early second-millennium Arabian astronomers estimated values in the neighbourhood of 23°33′ and a little later Ulugh Beg gave a value of 23°30′17″. Modern evidence indicates that the obliquity of the ecliptic has been decreasing by about 47″ per century and that the mean value in 1950 was 23°26′45″. Copernicus believed that it was decreasing by irregular jumps or even oscillated between increases and decreases in an irregular fashion.

Copernicus, without any real observational evidence, claimed that the sphere of the fixed stars was extraordinarily distant relative to the greatest planetary distance. This was a defence against the possible argument, later put forward by Tycho Brahe, that if the

Earth was in revolution around the Sun there should be a semi-annual parallax (which had not been detected) if the fixed stars were not extraordinarily distant. Digges, an English proponent of Copernicus's theory, in 1592 recognized that Copernicus had no need to continue the belief in a sphere of equidistant fixed stars. If the apparent diurnal motion of the stars is the consequence of the rotation of the Earth rather than the rotation of the celestial sphere, then the relatively fixed formation of the non-planetary stars does not require that they be equidistant.

Having worked out the general ideas outlined in *Commentariolus* in greater detail in what became *De revolutionibus*, Copernicus's long-delayed publication, he may have been reluctant to become involved in ecclesiastical controversy. But that seems unlikely as he was being urged to publish by various senior churchmen. Further, he dedicated his *magnum opus* when it was published to Pope Paul III. It is more likely that he was not fully satisfied with his analyses but did not see how he could further improve them.

In 1539 Rheticus, a young mathematician from Wittenberg, visited Copernicus in Frauenburg (now Frombork) and spent about two years studying the manuscript of *De revolutionibus* and discussing its contents with Copernicus. Rheticus finally persuaded Copernicus to let him arrange for publication. Rheticus took the manuscript in revised form to Nuremberg where he arranged for its printing. Being unable to remain to see it through the press, Rheticus handed over the supervision to Osiander who added a preface with which Copernicus would almost certainly not have agreed. Osiander's preface stated that the theories which were expounded were not to be taken as claims about what in fact occurred in nature but only as convenient modes of conception and analysis for purposes of predicting what could be observed to happen in nature. This may well have been a precautionary move on Osiander's part to shield the work from ecclesiastical condemnation.

Tycho Brahe (1546–1601) believed (i) that he could estimate the apparent size of the fixed stars (he was in fact estimating their apparent brightness), and (ii) that no star could be many hundred times more distant than Saturn for if they were more distant they must in view of their apparent size be of incredibly large real size (say as great as the alleged orbit of the Earth). Consequently, if the Earth was in orbit, there must be a semi-annual parallax or a displacement of the apparent direction of the brighter stars. He failed to detect such a parallax, which he assumed might be of several minutes of arc. Hence he rejected the Copernican hypotheses of the Earth's revolution. Brahe was wrong on almost every score. Stellar diameters cannot be measured by direct visual means even with the aid of the best telescopes, not available to Brahe; by saying 'direct visual means' I intend to exclude interferometry as used by Michelson and Hanbury Brown in the earlier and the latter part of the twentieth century. When semi-annual parallaxes were established in the 1830s, the largest (that of *alpha Centauri*) was found to be only 0".76. *Alpha Centauri* is more than 28 500 times more distant from the Sun than Saturn. Finally, the probable diameters of some stars, for example Betelgeuze, are greater than the diameter of the Earth's orbit. Despite this (as it turned out invalid) falsification by Brahe of Copernicus's hypothesis of the Earth's revolution the more venturesome astronomers within the next three-quarters of a century, such as Galileo and Kepler, were espousing Copernicus's hypothesis.

Brahe suggested an alternative ingenious planetary system. The Moon and the Sun were deemed to revolve around a stationary Earth and the star-like planets were deemed to revolve around the Sun in the order of distance, Mercury, Venus, Mars, Jupiter and Saturn.

Copernicus's star catalogue was generally based on Ptolemy's of almost fourteen centuries earlier. Apart from a few corrections of longitude or latitude based on Arabian or Copernicus's own observations, Copernicus's latitudes were the same as Ptolemy's and his longitudes were Ptolemy's corrected for a misunderstood rate of precession. Because he believed in a variable rate of precession, Copernicus decided not to measure longitudes from the First Point in Aries (the spring equinox). Instead he picked on a fixed star as the reference point – the most western in the horn of the Ram or our *gamma Arietis*. His listing of the constellations and the location of the stars within them are essentially the same as Ptolemy's. Many of the stars in Ptolemy's catalogue were too far south for Copernicus to observe them from Frauenburg.

Perhaps the most ironic circumstance in the development of Copernicus's planetary theory is to be found in the changes Kepler (1571–1636) had to introduce in order to get the Copernican theory, taken in general terms, to work more effectively in

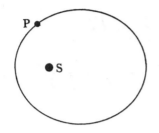

Fig. 13.6   *Kepler's first law: the orbit of a planet around the Sun is an ellipse with the Sun at one of the foci.*

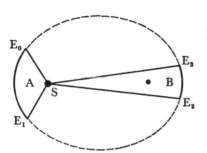

Fig. 13.7   *Kepler's second law: the radius vector from the Sun to the planet sweeps over equal areas in equal times.*

positional prediction than the Ptolemaic theories. Kepler abandoned Copernicus's circles upon circles and Copernicus's insistence on basic circular uniform motion. His painstaking attempts to find some pattern in Brahe's extensive data on the observed positions of Mars at given times led him to suppose that the orbit of a planet was an ellipse with the Sun at one of the foci (see Fig. 13.6) and that the radius vector from the Sun to the planet swept over equal areas in equal times (see Fig. 13.7). Thus the planetary orbit was seen by Kepler to be neither circular nor of uniform angular velocity. Even more ironic is that the other Keplerian focus was roughly equivalent to Ptolemy's equant, so abhorrent to Copernicus, the point from which angular velocity was held by Ptolemy to be uniform.

Ptolemy could not get a common centre in the Earth for all the other planetary orbits and Copernicus could not get it in the Sun. Kepler established that the Sun was a focus of all planetary ellipses (subject to some appreciable but small interplanetary perturbances).

When Newton found a dynamic explanation for Kepler's planetary laws, modern astronomy was launched but still had a long way to go.

CHINESE ASTRONOMY

Chinese astronomy dates from the thirteenth century BC and grew up largely independently of influences from the Near Eastern and Western astronomy which has been described in the preceding chapters. Nevertheless it reached considerable technical heights. The following account is not meant to be extensive but instead it aims to draw attention to similarities to and differences from the tradition discussed at length above. Chinese astronomy was already going into decline when the Jesuit missionaries, many of whom were interested in astronomy, arrived in China in the seventeenth century AD. The following summary is largely based on Needham's (1959) extensive discussion.

Whereas Near Eastern and Western astronomy used a pair of co-ordinates with an ecliptic orientation (the Arabs occasionally preferring a pair with an altazimuth orientation), the Chinese used a pair of equatorial-polar co-ordinates which have since been adopted by modern astronomy. The Chinese apparently knew of the Babylonian concept of 360 degrees in a circle but as they considered that the year contained $365\frac{1}{4}$ days they are said to have divided the circle into $365\frac{1}{4}$ degrees. The Chinese recognized more numerous and hence smaller constellations than did their contemporaries in the Near East and the West. By the middle of the first millennium AD some 50 constellations had been recognized and named in the West, whereas over 400 were recognized and named in China. Only about a dozen Chinese constellations had similar conformations to those in the West and even fewer had names with similar meanings.

Oracle-bones from the Shang dynasty, thirteenth to eleventh century BC, record dates of some eclipses and provide information implying a synodic month of 29.5 days and a year of 365.25 days. These bones were used in divination. Certain questions or pieces of information were incised or painted on them. Then upon being held over a fire cracks appeared in them which the augurs professed to be able to read for purposes of divination. After use these bones were buried in deep pits from which thousands have been recovered in modern times.

Some eclipse records other than the few on oracle-bones are available from 720 BC. With some serious gaps, over 900 solar eclipses and almost 600 lunar eclipses are recorded for the next 2000 years. There is only a moderate degree of reliability in the earlier of these eclipse dates when compared with modern calculations of them. Their reliability improves from the third century AD.

Early in the fourth century AD the precession of the equinoxes was discovered in China, although from a much earlier period a succession of abandoned 'pole-stars' had been replaced by a new one, a result of the unrecognized precession. It would seem that it was recognized as early as the first millennium BC that the celestial pole was shifting. By the fifth or sixth century AD the tropical year was deemed to be 365.242815 days and the sidereal year to be 365.257612 days; both are fair enough estimates but are sufficiently different from late Greek estimates as to be likely to be independent.

The inequality of the seasons seems not to have been discovered until the sixth century AD. The obliquity of the ecliptic was assessed as 23°50' in AD 200 and 23°32' in AD 1150. Planetary periods were markedly in error until the first century BC when the following values in days were asserted: Mercury, 115.9 (115.88), Venus, 584.1 (583.92), Mars, 780.5 (779.94), Jupiter, 398.7 (398.88), and Saturn, 337.9 (378.09), where the values in parentheses are mean modern values.

Quite early, perhaps in the first millennium BC, and almost certainly earlier than Meton, the Chinese established the near equivalence of nineteen years and 235 synodic months. Later they found they could establish a nearer equivalence by using 76 years and 940 months reduced by one day, probably independently of Kallippos's similar discovery.

The Chinese astronomers produced star maps, some drawn on paper and some embossed on metal sheets. The earliest surviving star maps date from about AD 900 but it is probable that the practice goes back to AD 300 or even earlier.

In addition to observing and recording eclipses, the Chinese from very early dates observed and recorded the dates of comets. For example, there are records of a very bright comet in 466 BC, in 239 BC and every 76 or 77 years thereafter; this almost certainly was Halley's comet, recorded in the West only much later. Meteor swarms are recorded from the seventh century BC and novae and supernovae (perhaps as far back as the thirteenth century BC). A notable observation of a supernova, said to be as bright as Venus for some weeks, was made in AD 1054; this is almost certainly the stellar explosion which produced the Crab Nebula, unrecorded in the Near East and the West. Sunspots were observed and reported in China from the first century BC, long before they were observed and reported in the Near East and the West. Hipparchos may have observed a nova in the second century BC but he may have mistaken a comet for a new star. Tycho Brahe and Kepler observed novae in 1572 and 1604 respectively and Galileo reported sunspots in 1610. The Chinese seem to have anticipated them by many centuries.

The Chinese had a set of 28 reference stars arranged roughly near the celestial equator. These stars, called *hsiu* or palaces of the Moon, were analogous in general terms to the 27 or 28 Indian *nakshatra* stars or brides of the Moon and to the Arabian *al-Manazil* or inns of the Moon. The similar generic name and similar numbers suggest a common origin. If this is so the Arabian scheme, being later, must be derivative. There are several important differences between the Indian *nakshatra* stars and the Chinese *hsiu* stars. Whereas the *nakshatra* stars are in general near to the ecliptic, the *hsiu* stars are near to the celestial equator. Both are unevenly spaced, but the spacing of opposing *hsiu* stars tends to be similar; sometimes the space between two *hsiu* stars on one side embraces two such spaces on the other side. Only nine *nakshatra* and *hsiu* stars are common, eleven are in the same western constellation and eight are in different constellations. If the two schemes had a common origin, they have obviously markedly diverged. Joseph Needham (1959) suggests that both had a common origin in Babylonia. An interesting and unique feature of the *hsiu* system was the designation of 28 circumpolar stars on approximately the same meridians as the *hsiu* stars. Thus even when a *hsiu* star was below the horizon its direction could be read from its circumpolar paranatellon (a star crossing the meridian at the same time).

By the second millennium AD, the Chinese developed highly sophisticated observational instrumentation. The early Chinese astronomers had used the vertical gnomon to measure, through the shadow cast, the altitude of the Sun at the solstices. The earliest observations of this sort may be dated from the late second millennium BC. Their use to determine the obliquity of the ecliptic may have been as early as about fourth century BC and was certainly no later than about AD 100. Much later great masonry structures were used in place of the vertical pole. They were in use early in the second millennium AD and may have originated in the late first millennium AD. They predated the earlier described large instruments in Arabia (Ulugh Beg), in Europe (Regiomontanus and Brahe) and in India (Jai Singh). Largeness in this case enabled greater positional accuracy.

Time-reckoning devices included (a) sundials, an extension of the simple gnomon, (b) the water-clock, and (c) the combustion-clock. There is evidence of sundials in the late first millennium BC, some with flat horizontal bases and some with a hemispherical 'base'. Later Chinese sundials had a flat base in the equatorial plane and with the gnomon pointing to the celestial North Pole in conformity with the equatorial-polar orientation of Chinese positional astronomy.

Water-clocks were either of the outflow type (most common in the Near East and possibly borrowed from there) or of the inflow type. In the former, time was measured by the decreasing height of water slowly leaking from a vessel, whereas in the latter, time was measured by the increasing height of water in a vessel fed by a slowly leaking source vessel. The Egyptians and perhaps the Babylonians recognized a problem with a single water-pot, whether outflow or inflow. When the leaking vessel is full the higher pressure results in a greater rate of leakage than when it is, say, only half full. The Egyptians corrected for this by using an unequal graduation of the points within the vessel to mark the hours. The Chinese like the Alexandrian Greeks used compensatory feeder vessels, the contents of which were kept at near constant volume. The Chinese likewise used floats in the final vessel to mark the time on a scale. Much later the flow of water was used to drive gear-chains connected with time indicators; this did not come until perhaps late in the first millennium AD or a little later.

The combustion-clocks, probably late first millennium AD, were metal or ceramic plates with a maze-like set of channels which were filled with some combustible material. When lit at one end the burning material marked off time as it turned one corner after another. Over short periods these combustion-clocks were more accurate than water-clocks, but burned out too quickly to be useful in measuring long periods.

The Chinese used a sighting tube in the first half of the first millennium AD to establish celestial directions. The observer aimed the tube through which he looked at some object and read off the position of the object on a pair of equatorial and polar scales. This sighting tube was later incorporated in some Chinese armillary spheres.

Armillary spheres came into use in China at about the beginning of the Christian era or a little later. They are so like the Greek instruments that it is difficult to believe that they were independent inventions. They began in quite simple forms but during the next millennium they became more complex, and were given an equatorial rather than an ecliptic orientation. These were almost certainly independent Chinese variations.

Some time in the first millennium AD the Chinese adopted the twelve signs of the zodiac. These were probably introduced by Buddhist monks from India. The Chinese, however, bestowed completely different animal names on them: rat, ox, tiger, hare, dragon, snake, horse, sheep, monkey, cock, dog and pig, which are the names bestowed on the Chinese cycle of twelve years.

For demonstration and catalogue purposes but not for use in observation, celestial globes with a polar orientation were constructed from at least as early as the second millennium AD. Star maps were inscribed on the surface of these globes.

Though there are many indications of some borrowing from the Near East and the West, Chinese astronomy seems to be predominantly an independent creation. After the arrival in the late seventeenth century of the Jesuit missionaries, many of whom were expert in astronomy, the by then ailing indigenous Chinese astronomy was progessively assimilated to the Western tradition.

POSSIBLE MEGALITHIC OBSERVATORIES

*I*t has often been argued, especially in quite recent times, that many of the megalithic monuments in the British Isles, in Brittany and elsewhere in Europe have astronomical significance. They are said to be sighting devices used to locate from the Sun's rising and setting points on the horizon the occasions of the solstices and equinoxes and to keep track of the Moon's shifting most northerly and most southerly rising and setting points, perhaps for the prediction of eclipses. Some, it is claimed, could also provide alignments to the rising and setting points of some of the brighter stars.

Later records suggest that such observations were being made in Mesopotamia at least as early as 3000 BC. Our earliest preserved records are the Venus tablets of Ammisaduga, apparently from the first half of the second millennium BC, though our copies, clearly full of errors, were made almost a thousand years later. The megalithic monuments under discussion were built in the neolithic and early bronze ages, say from about 2500 to 1500 BC. There is nothing to suggest that whatever astronomical activity was being engaged in, these remote outskirts of Europe provided any base for the scientific astronomy of the Near East and the later West and involving Mesopotamia, Greece, India, Arabia and finally a reawakened Latin Europe. The monuments could,

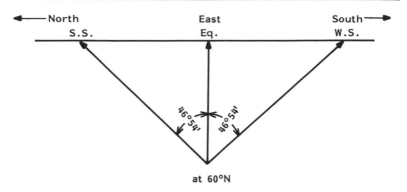

Fig. A2.1 *For an observer at terrestrial latitude 60°N the Sun will rise at the summer solstice 46°54' north of east, at the equinoxes due east and at the winter solstice 46°54' south of east.*

however, provide an indication of how far fairly unsophisticated cultures could go. Judging from associated pottery and other artefacts, and from burial customs, the monuments in Britain were built by successive pre-Celtic groups who had distinctive cultures and who had moved across from continental Europe or had assimilated practices from continental neighbours. Whereas the roughly contemporary priestly class in Mesopotamia was literate and numerate, these monument builders in Britain were almost certainly illiterate which makes the astronomical claims for them somewhat implausible. There is, however, some circumstantial evidence from the dimensions of the monuments that their builders had some mensurational and practical geometrical skills.

Where a celestial body rises or sets, an event on the horizon, depends on several conditions. First is its declination, that is, its placement north or south of the celestial equator. A celestial body with a declination of 0° from the celestial equator will rise due east and set due west; the Sun does so at the equinoxes. A celestial body with a declination north of the celestial equator will rise north of due east and set north of due west. Second is the terrestrial latitude of the observer. An observer at the terrestrial equator will see the Sun rise 23°27' north of east at the summer solstice if the horizon is 'level'; an observer at 30°N will see it rise 27°5' north of east, and an observer at 60°N will see it rise 46°54' north of east (see Fig. A2.1). Third, the values given are affected by the 'height' of the

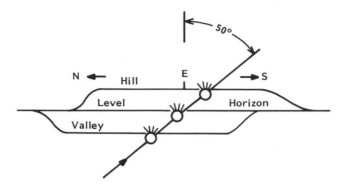

Fig. A2.2   *The effect of the height of the horizon on the point of sunrise for an observer 50°N of the terrestrial equator.*

horizon except when the terrestrial latitude of the observer is 0°. North of the terrestrial equator the body seems to rise on a southward slanting path, the further north the greater the slant. If a hill provides the horizon, the rising body will be caught later on its southward slanting path; if a valley provides the horizon, the rising body will be caught earlier on its slanting path (see Fig. A2.2)

Three less substantial conditions each deserve a brief mention. (i) As a result of atmospheric refraction for objects near the horizon curving the light-rays downwards, a celestial body which on a straight-line of sight is below the horizon is seen to be above the horizon. This upward displacement amounts to only a degree or so but it exaggerates the shift of apparent rising or setting points for observers well away from the terrestrial equator. (ii) Atmospheric absorption near the horizon renders all the fixed stars but Sirius and Canopus invisible even under good viewing conditions until they are a degree or more, depending on their magnitude, above the apparent horizon. The slanting path as seen by observers located away from the terrestrial equator would render the location of their actual rising or setting points more difficult to determine. (iii) The apparent direction of the celestial body will appear slightly different for observers at markedly different longitudinal positions on the Earth's surface. For naked-eye sighting devices, only lunar parallactic displacements are reliably discernible; other bodies are too distant for parallax to be detected without telescopic and other aids.

We also need to note a problem in specifying what is meant by the rising and setting of the Sun and the Moon, which have perceptible discs. Does the moment of rising and setting occur when only the top sliver of the disc can be seen above the horizon (the first and last flash respectively), when half the disc is visible, or when the whole disc sits on the horizon? For an observer at 45°N the rising Sun will have moved 30′ to the south between its first flash and its sitting as a whole on the horizon.

The megalithic monuments in Britain and Brittany consist of three types of structure, sometimes two or more in combination (see Burl 1976; Thom 1967 and 1971; Wood 1978). First there are rings of stones from, say, five to thirty or more in number. Second, there are rings of ditches and banks, sometimes associated with rings of stones, as at Avebury and Stonehenge. Third, there are pairs or rows of stones; sometimes a free-standing stone of this type is an 'outlier' associated with a ring of stones, as in the case of Long Meg and her daughters in Cumbria or the Heel Stone at Stonehenge.

Many of the stone monuments are in a ruinous or otherwise disturbed state: some stones have fallen and even if re-erected may not be restored exactly to their original position; stones not buried to the same depth have been differentially shifted by the flow of the topsoil, by trees growing up beside them and pushing them aside, by frost melting on the sunny side, softening the ground and so allowing displacement from the frozen side; and so on (see Atkinson 1956 and Burl 1976).

Even though the builders of these monuments were almost certainly illiterate there is some evidence to be found in the structures themselves that their builders were skilled applied geometers and had a concern with some numerical relationships.

Alexander Thom (1967, 1971, 1978), who has done most of the exact measurement in relation to these monuments, has surveyed with theodolite, compass and surveyor's chain several hundred of them. He maintains that in respect of diameters the builders used as units of length what he calls a megalithic yard (some 2.7 feet) and in respect of circumferences a megalithic rod (2.5 megalithic yards). He maintains that diameters are simple multiples of megalithic yards and circumferences are simple multiples of megalithic rods. His own evidence is strongest for Scotland and for Brittany (in the latter there are few rings) and weakest for England. Kendall

(1974) has argued on statistical grounds that the Thom samples are too small for a conclusive decision. Mackie (1977) on archaeological grounds comes to a more favourable conclusion. The issue of whether or not the monument builders had standard units of length does not bear much on Thom's other contention, supported by many others, that the monuments have intended solar and lunar alignments.

There are some 900 stone rings in Britain, some only remnants. Thom (1967) recognizes several type of such rings. The most numerous, about two-thirds, are near-perfect circles. These could have been laid out by means of a rope rotated around a central peg. Next are 'flattened' circles or segments of two circles of different radii put together; they constitute about one-sixth. Next are ellipses, about one-ninth. Least numerous are ovals or 'egg-shapes', about one-eighteenth. Ellipses can be traced out by means of a rope loop and two pegs separated by less than the length of the loop, and the 'egg-shaped' rings can be produced by tracing segments of circles of different radii from the apices of Pythagorean triangles. Whether neolithic men in Britain so produced these rings we can only conjecture. But however they generated these shapes they appear to have been skilled practical geometers, if we allow for the possible distortion of what may have been their original arrangements as a consequence of the possible distorting factors mentioned above.

It is not the stone rings that most strongly suggest astronomical alignments. Indeed such rings tend to have too many alignments. Consider a circle of 30 stones such as Stonehenge. Each stone has 29 alignments with its neighbours so there is a total of 870 alignments, a rather excessive number for major astronomical events. If the 30 stones in the circle were evenly spaced the alignments would be separated by about 25′ of arc. Further, even stones on opposite sides of a large ring are not far enough apart. To use a pair of stones as a backsight and a foresight for the location of some point on the horizon they need to be separated by several kilometres if they are to define with any precision a point on the horizon. At Stonehenge opposite major stones are only 33 metres apart. This is a problem for the suggestion that Stonehenge is a solar and a lunar observatory. Sometimes there is an outlying stone such as the Heel Stone at Stonehenge (see Atkinson 1956) or Long Meg in relation to 'her daughters' (see

Burl 1976) which is thought to be a foresight viewed from say the centre of the circle. For instance, the Heel Stone has been thought to be a foresight for an observer at the centre of Stonehenge wishing to identify the sunrise at the summer solstice. There are three problems here. First, what is the centre of Stonehenge? The centre of the bank and Aubrey holes and the centre of the Sarsen circle differ by about a metre. Second, how would the observer know that he was at the centre? There is no evidence of a stone or a post to mark it. Third, the 'centre', wherever it be, and the Heel Stone are not so far apart (less than 90 metres) that a precise alignment on the horizon, say within a degree, could be established. There is the further problem already mentioned. How does one determine the point of sunrise on the horizon? As we have noted, we do not know which point on the horizon may have been taken. Perhaps the first flash at sunrise or the last flash at sunset are the most plausible.

The proposals for Stonehenge as a solar and lunar observatory and as an eclipse computer will be examined later and given a low probability or plausibility rating. The better cases are to be found with the relatively isolated free-standing stones or menhirs in pairs, in triples and so on and sometimes forming alignments with natural objects such as clefts in or prominences on distant hills; sometimes a pair of not widely separated menhirs indicate the direction in which the foresight is to be looked for. There are many such examples in Scotland, northern England and Brittany.

There are several passage tombs such as that at New Grange, north of Dublin in the Boyne Valley, which allow the rising Sun for a few days around the winter solstice to illuminate the inner recesses of the tomb (see Herity 1974). This inner illumination did not pinpoint the day of the winter solstice but bracketed it. Perhaps the winter solstice was being thought of as an occasion of resurrection, as was the later Roman *dies natalis solis invicti*, the birthday of the unconquered Sun, which the Christians adopted as Christ's birthday.

A few more general problems may be mentioned. As already indicated the rings of stone have an embarrassing number of alignments which are far in excess of likely astronomical alignments of interest. Further, many of the simpler monuments seem to have no likely astronomical alignments. A religious or ceremonial significance seems more likely for the rings, though some

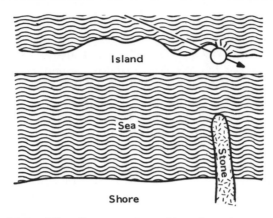

Fig. A2.3   *The alignment claimed by Thom for a winter solstice at Kintraw.*

of the festivals celebrated in them may have had to do with equinoxes and solstices. A few examples of the more convincing solar alignments will be given. At Ballochroy on the peninsula of Kintyre in Scotland, the centre member of a set of three menhirs aligns with a slight notch in a sloping hillside on the island of Jura, some 30 km away, where there would be seen a last flash of sunlight as the Sun was setting at the summer solstice. Some 50 km north, at Kintraw, there is an apparent viewing platform below which was a single menhir providing an alignment to another notch in a slope on Jura, some 43 km away, from which there would be a last flash of sunlight as the Sun was setting at the winter solstice (see Fig. A2.3). Recently doubts have been cast on the astronomical utility of these two alignments (Patrick 1981); the second is said to be blocked by near topographical features and the first to be obscured by atmospheric conditions.

The west and north coasts of Scotland and the northwest of England provide the most numerous examples of these solar alignments. As a rule, however, one site provides only one usable alignment, whereas it would be more convenient, for use in an elaborate calendrical scheme, if several or all of the necessary alignments were provided on the one site or at least close neighbouring sites. There are few cases of plausible multi-alignments at one site in Britain. Of the alleged solar alignments the most numerous sites relate to sunrise or sunset at the summer or winter solstices, that is declinations to which allowances are added for the terrestrial latitude of the site. The declinations average about 23°58′N or S (correct for the obliquity of the ecliptic in about 2050 BC) with a mean deviation of about 18′. Alignments to the equinoxes are much less numerous; they have a mean value of 0°21′N, with a mean deviation of about 26′. Because the Sun is moving north-south more rapidly at the equinoxes than at the solstices, the former should be more accurately located than the latter. However, the location of the direction of the equinoxes depends upon having a good compass: perhaps the best compass neolithic man had was the direction of sunrise and sunset at what were taken to be the equinoxes, which he probably assumed to occur on dates midway between the solstices and which are about two days away from the true equinoxes as a result of the inequality of the seasons. In our time the intervals between summer and winter solstice and between winter and summer solstice are about 183 days 6 hours and 181 days 22 hours respectively. In, say, 2000 BC these intervals would have been different but the principle would have been the same.

Almost as numerous as alleged equinoctial alignments are two sets of solar horizon alignments corresponding to points midway on a rough day count between solstices and equinoxes. These are dates early in our May, August, November and February. They correspond roughly to the medieval feast days of May Day (1 May), Lammas (1 August), Martinmas (11 November) and Candlemas (2 February). The spring equinox as assessed in early medieval times became the date of incarnation of Jesus and the winter solstice became his birthday; the autumn equinox became the date of John the Baptist's conception and the summer solstice became his birthday. It would, therefore, not be surprising if four neolithic intermediate dates were also later given Christian religious festival status.

Slightly less frequent were alignments to eight other inter-mediate dates, giving a total of sixteen calendar dates some 22, 23 or 24 days apart. This suggests a calendar dividing the year into sixteen segments each shorter than the lunar synodic month (see Thom 1967).

The changing direction of sunrise and sunset follows a regular, simple pattern which is easily discerned. The declination varied in, say, 2000 BC in approximately half a year from almost 24°N to almost 24°S; further, while the time intervals between inter-mediate directions on the horizon are not quite constant they are nearly so. By contrast the Moon presents a much more complex pattern. The Moon zigzags at an angle of about 5°9' above and below the ecliptic or apparent path of the Sun twice in about 27.212 days, once going north and once going south. The points where the Moon crosses the ecliptic are called nodes. Just as the equinoxes (the Sun's crossings of the celestial equator) precess, i.e. move westward, so do the lunar nodes; the former shift over 360° in 25 800 years, the latter do so in about 18.6 years. Thus if at a given time the maximum declination of the Moon is about 5°9' less than the maximum declination of the Sun (that is, about 18°49' as compared with about 23°58' from the celestial equator, to use a value appropriate for about 2050 BC), then 9.3 years later it would be about 5°9' more (that is, about 29°7'). Thus the Moon swings between ± 29°7' (called by Thom its maximum standstill) and ± 18'49' (the minimum standstill).

The Moon has another, though quite small, variation called the perturbation. When the Sun, the Moon and the Earth are in line, the Moon's inclination to the ecliptic is increased by about 9'. Thus the Moon's declination has three components: ε, the obliquity of the ecliptic (about 23°27' today but about 23°58' in say 2000 BC), ι, the inclination of the orbit of the Moon to the ecliptic (about 5°9'), and δ, the perturbation (about 9'), the last barely detectable with primitive sighting devices.

Thom (1971) has analysed 40 alignments of menhirs which seemed to him to have lunar alignments and for which he could establish precise declinations. His analysis was aimed at establishing precisely the four unknowns, ε, ι, δ, s (the last being the semi-diameter of the Moon as seen from the Earth). He used simultaneous equations with three or four unknowns. The values yielded are 23°53'20", 5°5'52", 9'23" and 15'55" respectively, which are surprisingly good for an epoch in say the early second millennium BC. These values may be as good as they are because they are means of a selection of promising instances.

Amongst the probable lunar alignments are the following. At Ballinaby in the Hebrides, a menhir 5 m high lines up with a sloping hillside 2 km away, which the upper edge of the setting Moon would graze when the declination of the Moon was $+\varepsilon+\iota+\delta$, that is, at a major northern standstill. From an egg-shaped stone ring on Dartmoor, the lower edge of the rising Moon would graze a ridge on a hillside 6.5 km away when the declination of the Moon was $-\varepsilon$ $-\iota$ $-\delta$, that is, at a major southern standstill. At Temple Wood, near the solar observatory at Kintraw, there is a complex set of standing stones which give alignments for the setting Moon at a notch 2 km away for declinations of $+(\varepsilon+\iota)$; $+(\varepsilon+\iota+\delta+s)$ and $(\varepsilon+\iota+\delta)$, and to a notch 6.3 km away when the declination of the Moon is $-(\varepsilon+\iota)$. The actual usability of some of these alignments has been questioned (Ruggles and Whittle 1981).

Perhaps the most remarkable probable lunar observatory was in Brittany near Carnac. A gigantic menhir, now fallen and broken, originally stood some 19 m above ground level. Its top was visible in the relatively flat terrain from points 15 km distant or more. It could have been used as a common foresight aligned with less impressive backsights circling it. There could have been, in terms of the terrain, backsights enabling the determination of the major and minor standstills, that is, when the Moon's declination was $+\varepsilon+\iota$; $-\varepsilon+\iota$; $+\varepsilon-\iota$; $-\varepsilon-\iota$ when rising or setting. Foresights for only three of these eight alignments (four rising and four setting Moon positions) have been found. The others may not have been preserved or if preserved may not have been discovered yet; the terrain in places is now heavily overgrown by trees and shrubs where earlier the view may have been clear.

There are a few other moderately frequent alignments which cannot be either solar or lunar. They could be stellar, but the case for them being so is not strong as they are not alignments to the brightest stars. Perhaps they throw some doubt on other alleged astronomical alignments.

It is possible, indeed probable, that the builders of these megalithic monuments were concerned with certain seasonal events, for example, midwinter and midsummer, spring and autumn, which they celebrated within their ring structures. In order to establish reasonably precise dates for these events they may have erected sets of stones by means of which sunrise or sunset on these occasions could be roughly established. It is a matter for puzzlement that they should have been concerned with the 'swings', maximum and minimum, of the Moon. One possibility has to do with eclipse warnings which will be dealt with later.

A brief comment on Stonehenge as a possible observatory

seems required because claims for its being so have often been made (see Krupp 1979). Perhaps the earliest such claim is that the summer solstice sunrise may be seen by an observer at the centre of the monument to occur at the top of the Heel Stone. There are several problems here, as we have seen. Where precisely is the centre? What is sunrise (the first flash, a bisected solar disc, the full disc sitting on top of the horizon)? Further, in say 2500 or 2000 BC, the Sun would have risen to the north of the Heel Stone and would have been above it as viewed from the centre of the bank and Aubrey holes some little time after sunrise at the summer solstice. Allowance must be made here for the decrease in the obliquity of the ecliptic to the celestial equator, which can be calculated, and the height of the soil within the bank at Stonehenge, which can only be guessed.

In 1965 Hawkins reported a study of Stonehenge alignments. He took a diagram of the locations of the Stonehenge stones standing and of the probable locations of those that have fallen or been removed and calculated alignments to the horizon (as suggested by the contour lines on ordinance maps). He found a large number of apparently significant solar and lunar alignments between pairs of stones or gaps between stones. There were some serious deficiencies in these findings, as was pointed out by Atkinson (1966a, 1966b). First, the specified alignments were less precise than suggested: for example the positions of two stones on a diagram could not define as precise an alignment as would a careful survey such as subsequently made by Thom (1974, 1975). Second, lines of sight through gaps between pairs of stones, without allowing for shifts of head positions, were far from precise. Most importantly, Hawkins's alignments involved backsights and foresights so close together as to allow only a rough bearing on the horizon. Later Atkinson (1974, 1975) admitted that he may have over-reacted at first to Hawkins's contentions.

The careful survey of Stonehenge by Thom (1975) has confirmed some of Hawkins's alignments though the spacing of the backsights and the foresights are still too short for a reasonable degree of precision for celestial events on the horizon. Thom has found some possible foresights in the Stonehenge environment distant enough to provide the necessary precision. It has not been established, however, that the alleged foresights which are man-made have the relevant prehistoric origins.

In brief, it may be doubted that Stonehenge was a solar and lunar observatory as Hawkins maintained and as Thom is inclined to agree; this doubt, however, is different from a belief that it may have been a 'temple' for the celebration of equinoxes and solstices and so had rough alignments to sunrise and sunset in the neighbourhood, say the week, of these occasions. There is a much better case for other British monuments being more precise solar or lunar observatories. This case is strengthened by evidence presented in the name of archaeo-astronomy (see Krupp 1979 and Aveni 1977) for later astronomical observations by more or less sophisticated peoples in Central America (the Mayas) and in mid-western U.S.A. (the Plains Indians).

Let us now turn to the issue of possible megalithic British eclipse prediction. Stonehenge will be my first concern. Hawkins argued that the 56 Aubrey holes which are fairly evenly spaced in a near circle just inside the bank at Stonehenge can be used as a computer for predicting eclipses. A lunar eclipse occurs when the Moon is full and is near a node (one of the two points where the Moon's apparent path crosses the ecliptic). A solar eclipse occurs when the Moon is in conjunction with the Sun and is near a node. The nearness to a node is within about ± 6° for a total lunar eclipse and within about ± 12° for a partial eclipse, whereas for a partial solar eclipse the nearness may be up to about ± 15°. A lunar eclipse may be seen from any part of the Earth where the eclipsed Moon is above the horizon (at night). Thus from a given site about one in two lunar eclipses may be observed (subject, of course, to reasonable viewing conditions). Solar eclipses, total and partial, can be seen only from relatively narrow curved strips on the sunlit side of the Earth. Only one in about six or seven solar eclipses can be seen from a given site, and then only as partial eclipses on most occasions.

Hawkins suggests the following use of the 56 Aubrey holes at Stonehenge as an eclipse computer. A full Moon rising near midwinter with a declination of about 24° (in megalithic times) would almost certainly have been eclipsed. At Stonehenge it would have risen over the Heel Stone. If on this occasion one of six marker stones was placed at Aubrey hole 56 just to the left of the alignment to the Heel Stone, the eclipse would be recorded.

Five other markers would have to have been placed at Aubrey holes 46, 37, 28, 18 and 9, that is spaced 10, 8, 9, 10, 9 and 9 holes apart. All six stones, according to Hawkins's theory, would have to be moved forward one hole per year. When the marker that had been at hole 56 had been shifted to hole 51, there was another eclipse warning. The crucial holes are 5, 56 and 51, which the circulating markers reach after five, four or five years. Lunar eclipses occur in series at intervals of six synodic months for seven or eight instances, each such series being separated by five months, about half of which would be seen from a given site. Hawkins's computer would predict only about one in eight or ten of all eclipses – a rather poor achievement.

Hoyle (1972) suggested the use of four markers shifted around the Aubrey holes. One marker, indicating the Sun's eastward progress, was to be moved anticlockwise by two Aubrey holes every thirteen days; this would produce an error of about 1° per annum but this could be prevented from cumulating over the years by a midsummer resetting. A second marker was to be moved anticlockwise by two Aubrey holes per day; it would be nearly opposite to the Sun's marker every month and could be reset. Finally a pair of markers opposite to each other were to be moved anticlockwise by three Aubrey holes per year to mark the positions of the ascending and descending lunar nodes, N and N'. When the Moon marker was at N and the Sun marker was within ± 15° of N, a solar eclipse was possible. When the Moon marker was at N and the Sun marker was within ± 10° of N' (or vice versa), a lunar eclipse was possible. As stated above, only one out of six or seven solar eclipses is visible from a given site whereas about one in two lunar eclipses is visible from a given site. Whereas Hoyle's method of using the Aubrey holes for eclipse prediction covers a substantially greater number of possible eclipses than does Hawkins's, it is so complex as to be implausible for the period and people concerned. While early untutored sky-watchers might discover that the apparent paths of the Sun and the Moon were usually different but crossed twice in about 27.2 days, it is straining credulity to suggest that they recognized these crossing points to have a direction which moved east to west through a full circle in 18.6 years. Finally, how could they have judged when the Sun marker was within ± 10° or ± 15° of N or N'? The average spacing of the centres of the Aubrey holes is 6°24' and not 5° which would yield multiples such as 10° and 15°.

Colton and Martin (1967, 1969) suggested a very simple

method to predict eclipses by use of the Aubrey holes. The circle of holes may have been used as a protractor by an observer standing in the centre (an apparently unmarked point) to establish whether or not the rising Moon is diametrically opposite the setting Sun. If the Moon and the Sun were at their rising points opposite to each other and the Moon rose up to 15 to 30 minutes before the Sun set, there would be an eclipse during the ensuing night. If the interval between moonrise and sunset were longer the eclipse that would follow would occur after moonset for that site. If the Moon rose after sunset, the eclipse had already occurred. Apart from the slight uncertainty in the time interval this simple method not only enables the prediction of an eclipse but also the prediction of whether or not it would be visible from that site.

Thom proposed an eclipse predictor for other monuments. He maintains that the value δ, the perturbation in the lunar declination, could be used for eclipse prediction. An eclipse will occur only when δ is at its maximum. However, according to Thom's contentions, megalithic observers could assess δ only at a maximum or minimum lunar standstill, which are about nine years apart. In nine years there would be about fifteen or sixteen lunar eclipses of which about half would be visible from the given site. This, therefore, is a poor method of eclipse prediction. The case for systematic megalithic eclipse prediction does not seem to be strong. Hawkins's and Thom's proposals miss too many eclipses, and Hoyle's is so complex and calls for so high a degree of sophistication as to be improbable. Colton and Martin's proposal is credible, although protractor markings consisting of white clay patches 75 to 175 cm in diameter and over 20 m distant at ground level inside a ditch surrounded by a 2 m high bank would not be readily visible at sunrise and sunset.

In summary it may be said that though some of these megalithic monuments had approximate alignments to the rising and setting points on the horizon of the Sun at solstices and equinoxes and of the Moon at major and minor standstills and may have been devoted to the celebration of seasonal festivals, few enabled the pinpointing to a day or two of these celestial events. It would, of course, be enough to locate within a span of a few days the occurrence of a solstice or an equinox in order to conduct seasonal festivals at an appropriate time. These ancient peoples may well have noticed the differing swings of the Moon between major and minor standstills and have attached some significance to the phenomena but it may be doubted that they could use this information to predict eclipses. The Aubrey holes at Stonehenge would have been poor aids in observing solar and lunar rising and setting points. They were at ground level and a bank beyond them probably cut off the horizon. Broad stones and broad gaps between stones provide poor sights. While it is quite proper to call Jai Singh's naked-eye sets of sighting devices in Delhi, Jaipur and Benares 'observatories', this seems an exaggerated claim for the megalithic monuments in Britain and Brittany.

GLOSSARY

ALTITUDE the angular height of a celestial body above the horizon as measured along a line from the body perpendicular to the horizon.

ANOMALISTIC MONTH the interval between successive lunar perigees – 27.55455 days.

ANOMALISTIC YEAR interval between successive passages of the Earth through perihelion (or aphelion) – 365.25964 days.

ANOMALOUS APPARENT PLANETARY MOTION variation in the apparent velocity of 'planets'.

APHELION when a planet is furthest from the Sun.

APOGEE furthest from the Earth, now applied only to the Moon.

ASTERISM a close set of stars less extensive than a constellation, e.g. the Pleiades and the Hyades; these are usually stellar clusters.

AZIMUTH the angular separation of a point on the horizon from due north usually measured eastward.

CELESTIAL EQUATOR the plane of the Earth's rotation, at right angles to the axis of rotation; *see* celestial poles.

CELESTIAL POLES the points north and south through which the Earth's axis of rotation passes.

CONJUNCTION when two or more 'planets' (including the Sun and the Moon) have the same geocentric longitude or the same Right Ascension. Superior planets in the modern sense are in conjunction with the Sun when they are on the opposite side of it from the Earth. Inferior planets have two conjunctions with the Sun: one when on the opposite side of the Sun and the other when between the Earth and the Sun, superior and inferior conjunction respectively.

CONSTELLATION a collection of stars most of which form some recognizable pattern. Until recent times a distinction was made between the main pattern (the *imago*) and surrounding stars. In recent times the constellations include all stars within fixed straight-line boundaries.

CO-ORDINATE SYSTEMS   there are three main pairs of co-ordinates for the positional location of celestial bodies: (a) the altitude-azimuth pair which uses the horizon as its reference plane, (b) the longitude-latitude pair which uses the ecliptic, and (c) the right ascension-declination pair which uses the equator.

CRANK MECHANISM   a geometrical device used by Ptolemy in which the centre of a deferent (q.v.) moved around a small circle or sub-deferent.

DECAN   derived from the Greek name used for each of an Egyptian set of stars, the rising of which was used to mark the passage of the hours at night. As the stars rise four minutes later night by night a given decan was replaced after ten days by its predecessor to signal a given hour. Later the term was used in astrology to designate a third of a sign of the zodiac.

DECLINATION   the angular separation of a celestial body north (+) or south (−) from the celestial equator.

DEFERENT   the basic cycle in Hipparchan-Ptolemaic and Copernican planetary theories on the circumference of which the centre of an epicycle or set of epicycles is borne.

DIARY (BABYLONIAN TABLE)   a record, usually extending over half a year, giving the days of the month on which important celestial phenomena occurred and including reports of some civil events.

DIRECT APPARENT PLANETARY MOTION   the apparent eastward motion of the 'planets' amongst the fixed stars.

DRACONIC MONTH   interval between the Moon's passage through corresponding nodes − 27.21222 days.

ECLIPSE CYCLE   a recurring sequence of solar and lunar eclipses involving a period of 18 years 11 days after which the Sun, the Moon and the lunar nodes (q.v.) return to the same positions. Halley called it the Saros cycle mistaking the meaning of an Akkadian word.

ECLIPTIC   the apparent path of the Sun amongst the fixed stars; the plane of the Earth's orbit around the Sun.

EPHEMERIS (*pl.* EPHEMERIDES)   a table showing the expected celestial positions of celestial bodies at specified times.

EPICYCLE   a subsidiary circle (or set of circles) borne on a deferent, in Hipparchan-Ptolemaic and Copernican theories used to account for apparent planetary motion.

EQUANT   a medieval term used for a concept introduced by Ptolemy: the centre of the epicycle was deemed to move with uniform angular velocity not in relation to the centre of the deferent or to the offset Earth, but to a point offset on the other side of the centre of the deferent from the Earth.

EQUINOXES   the Sun passes north of the celestial equator at northern springtime and south in autumn; these crossing points are the equinoxes when the intervals between sunrise and sunset (day) and sunset and sunrise (night) are equal.

FIXED STAR   until recent times an easy distinction could be made between the wandering celestial bodies (planets) and the majority of stars which seemed to pass across the sky in fixed formation; we now know that most so-called fixed stars have slow proper motions.

GEOCENTRIC (GEOFOCAL)   the Earth regarded as occupying the centre (or focus).

GOAL TEXT   a type of Babylonian table making use of resonance periods (q.v.) in which past observed cycles were used to predict future planetary phenomena, such as disappearances, reappearances, conjunctions and so on.

HELIOCENTRIC (HELIOFOCAL)   the Sun regarded as occupying the centre (or focus).

HOMOCENTRIC SPHERES   a type of geometric analysis of apparent planetary motions introduced by Eudoxos and separately elaborated by Kallippos and Aristotle: spheres were deemed to be rotating within spheres on displaced axes.

LATITUDE   the angular separation of a celestial body north or south from the ecliptic.

LINE OF APSIDES   the line from the planetary, in the ancient sense, perigee to the apogee; in the modern sense, except in the case of the Moon, the line from perihelion to aphelion. The line slowly rotates at a rate peculiar to each 'planet'.

LONGITUDE   the angular separation of a celestial body from the point of the spring equinox (first point of Aries) measured along the ecliptic.

LUNAR INCLINATION (ι)   the angular displacement of the plane of the Moon's orbit from the plane of the Earth's orbit, approx. 5°.

LUNAR NODES   the points where the Moon's orbit cuts the plane of the Earth's orbit: the Moon is going north at the ascending node and south at the descending node.

MAGNITUDE OF STARS   the apparent brightness of a star; in the original scheme, 1 was applied to the brightest stars and 6 to those just visible to the naked eye.

MERIDIAN   a great circle passing through the celestial poles and cutting the celestial equator at right angles.

MONTH   a lunar period; see anomalistic month, draconic month, sidereal month and synodic month.

OBLIQUITY OF THE ECLIPTIC (ε)   the angular separation of the ecliptic (q.v.) from the celestial equator; it is variable and now has a mean value of approx. 23°27′ and is declining by 0.47″ annually.

OPPOSITION   when two or more 'planets' (including the Sun and the Moon) are separated by 180° in longitude or in Right Ascension. Mercury and Venus are never in opposition to the Sun.

PERIGEE   nearest to the Earth; term now applied only to the Moon or artificial satellites but once applied to all so-called 'planets'.

PERIHELION   when a planet is nearest to the Sun.

PLANET   from a geocentric viewpoint the Sun, the Moon, Mercury, Venus, Mars, Jupiter and Saturn wander amongst the fixed stars in a band called the zodiac; the Greeks called them *planetai*, wanderers. When referring to these seven apparent wanderers I usually write 'planets' or 'planets in the ancient sense'. In post-Copernican terminology the Sun is not a planet nor is the Moon (now regarded as a satellite) whereas the Earth has joined the five star-like planets as an orbiter around the Sun; the telescopic objects Uranus, Neptune and Pluto have been added in recent times.

PLANETARY STATION   point at which a planet appears to change from direct to retrograde motion and vice versa.

PRECESSION OF EQUINOXES   a westward motion of the nodes of the ecliptic (the equinoxes) along the celestial equator at the rate of 50.2″ per annum; it results in the zodiacal constellations slipping eastward out of the zodiacal signs and the tropical year being slightly shorter than the sidereal year.

QUADRATURE   when the angle Sun-planet (or Moon-Earth) is 90°.

RESONANCE PERIOD   a Babylonian concept recognizing that

planetary phenomena relative to the Sun (disappearances, reappearances, conjunctions, etc.) occur in cycles peculiar to each planet; thus the resonance period of Venus is a few days over eight years.

RETROGRADE APPARENT PLANETARY MOTION   phases of east to west apparent motion of the star-like 'planets'.

REVOLUTION   the motion of a subsidiary body around a primary body, e.g. the motion of the Moon around the Earth or of the Earth around the Sun.

RIGHT ASCENSION   the position of a celestial body relative to the spring equinox (first point in Aries) measured along the celestial equator, now usually in hours (one hour equals 15 degrees) and minutes.

ROTATION   the motion of a celestial body around its own axis, e.g. the Earth in a day; contrasted with revolution.

SAROS CYCLE   the period of 223 synodic months over which eclipses repeat themselves, discovered by the Babylonians and to which Halley erroneously attached their word *saros*.

SIDEREAL MONTH   the interval between the successive conjunctions of the Moon with some fixed star – 27.32166 days.

SIDEREAL PERIOD   the interval between the return to a given fixed star of an apparently moving celestial body, e.g. the sidereal year of the Sun, or the sidereal period of Venus.

SIDEREAL YEAR   the time taken by the Earth to make one complete circuit of the stars as seen from the Sun – 365.25636 days.

SOLSTICES   the occasions and the positions when the Sun is furthest north or south of the celestial equator; for a day or two the Sun seems to stand still, *sol stetit*.

SYNODIC MONTH   the interval between successive conjunctions between the Sun and the Moon – 29.53059 is the mean value.

SYNODIC PERIOD   the interval between successive conjunctions of a planet or the Moon with the Sun.

SYZYGY   opposition or conjunction of Sun and Moon.

TREPIDATION OF PRECESSION OF THE EQUINOXES   many Arabian astronomers supported the notion, attributed to unnamed Greeks, that the precession of the equinoxes increased and decreased in some cycle such as six or eight years: this seems to have been based on some erroneous estimates of precession, especially Ptolemy's.

TROPICAL YEAR   the interval between the successive passage of the Earth through the spring equinox – 365.24219 days, decreasing slowly over time.

YEAR   a period based on the apparent motion of the Sun. Three such periods are recognized: (a) the tropical year, the period from one spring equinox to the next, now 365.24219 days; (b) the sidereal year, the period between successive conjunctions of the Sun with some fixed star, 365.25636 days; (c) the anomalistic year, the period between successive perihelia (in ancient terms perigees), 365.26964 days. These periods are slowly changing.

ZENITH   the celestial position immediately overhead for the observer; the position under the observer's feet and 180° from the zenith is the nadir.

ZODIAC   a band some 8° wide centred on the ecliptic, called by the Greeks *Kyklos zodiakos*, the circle of the beasts, so named for their zodiacal constellations.

ZODIACAL CONSTELLATIONS   in Babylonia twelve constellations were recognized as marking in a general way the twelve lunar months and the apparent path of the Sun; these with possibly two name changes were adopted by the Greeks.

ZODIACAL PERIOD   interval between successive returns of a planet to a given point of the zodiac, as seen from the earth.

ZODIACAL SIGNS   early in the second half of the first millennium BC the Babylonians substituted 30 degree segments of the ecliptic for the earlier zodiacal constellations which were unequal in longitude and transferred to them the names of the constellation contained in them.

BIBLIOGRAPHY

Aaboe, A. (1958). Babylonian planetary theories. *Centaurus*, 5, 209–27.
—— (1964). On period relations in Babylonian astronomy. *Centaurus*, 10, 213–31.
—— (1974). Scientific astronomy in antiquity. In Hodson 1974.
Abbud, F. (1962). The planetary theory of Ibn al-Shatir: reduction of the geometric models to numerical tables. *Isis*, 53, 492–9.
Allen, R. H. [1899] (1963). *Star names: their lore and meaning*. New York, Dover. Corrected reprint of original edition, *Star-names and their meanings*.
Armitage, A. (1959). *Copernicus, the founder of modern astronomy*. New York and London, Thomas Yoseloff.
Atkinson, R. J. C. (1956). *Stonehenge*. London, Hamish Hamilton. Reprinted by Penguin Books, 1960.
—— (1966a). Decoder misled. *Nature*, 210, 1302.
—— (1966b). Moonshine on Stonehenge. *Antiquity*, 40, 212–16.
—— (1974). Neolithic science and technology. In Hodson 1974.
—— (1975). Megalithic astronomy – a prehistorian's comments. *J. Hist. Astr.*, 6, 42–53.
Aveni, F. A. (ed.) (1977). *Native American astronomy*. Austin, University of Texas Press.
Bienkowska, B. (ed.) (1973). *The scientific world of Copernicus*. English translation from Polish by C. Cenkalsky. Dordrecht, D. Reidel.
Boll, F. (1901). Die Sternkataloge des Hipparch und des Ptolemaios. *Bibliotheca mathematica*, 3 Folge, Bd. 2.
Brown, P. L. (1979). *Megaliths and masterminds*. London, Robert Hale.
Burl, A. (1976). *The stone circles of the British Isles*. New Haven, Yale University Press.
Clark, W. E. (1937). Science. In G. T. Barratt (ed.) *The legacy of India*. Oxford, Clarendon Press.
Colton, R. and Martin, R. L. (1967). Eclipse cycles and eclipses at Stonehenge. *Nature*, 213, 476–8.

—— (1969). Eclipse predictions at Stonehenge. *Nature*, 221, 1011–12.

Copernicus, N. *Commentariolus*. English translation in E. Rosen (1959), *Three Copernican treatises*. New York, Dover.

—— (1543). *De revolutionibus orbium coelestium*. Nuremberg. There are three English translations by: (i) C. G. Wallis (1952). Encyclopaedia Britannica; (ii) A. M. Duncan (1976). Newton Abbot, David and Charles; (iii) E. Rosen (1978). London, Macmillan. All are deemed by critics to be defective in one way or another but it seems that Rosen improved on Duncan and Duncan on Wallis.

Crombie, A. C. (1953). *Robert Grosseteste and the origins of experimental science, 1100–1700*. Oxford, Clarendon Press.

Cumont, F. V. M. (1912). *Astrology and religion among the Greeks and Romans*. New York, Putnam.

de Saussure, L. (1919). Le zodiaque lunaire asiatique. *Arch. sci. phys. nat.*, 1, 105–26.

—— (1919–20). Le systeme astronomique des chinois. *Arch. sci. phys. nat.*, 1, 186–216 and 561–88, 2, 214–31 and 325–50.

—— (1923). Origene babylonienne de l'astronomie chinois. *Arch. sci. phys. nat.*, 5, 5–18.

Dicks, D. R. (1954). Ancient astronomical instruments. *J. Br. Astr. Ass.*, 64, 77–85.

—— (1970). *Early Greek astronomy to Aristotle*. London, Thames and Hudson.

Dreyer, J. L. E. [1905] (1953). *A history of astronomy from Thales to Kepler*. New York, Dover. Originally published as *A history of planetary theory from Thales to Kepler* by Cambridge University Press.

—— (1917–18). On the origins of Ptolemy's catalogue of stars. *Royal Astronomical Society Monthly Notices*, 528–39; 343–9.

Duncan, A. M. (1976). Translation of Copernicus, *De revolutionibus*. Newton Abbot, David and Charles.

The encyclopaedia of Islam. 1st edn, M. T. Houtsman *et al.* (eds), 4 vols, 1913–34. 2nd edn, H. A. R. Gibb *et al.* (eds), Vol. I, 1960, B. Lewis *et al.* (eds), Vol. II, 1965, Vol. III, 1971, incomplete. Leiden, E. J. Brill; London, Lusac and Company.

Geminos. *Eisagoge eis ta phainomena*. Greek text edited and translated into German by C. Manitius (1898). Leipzig, Teubner.

Gillespie, C. C. (ed. in chief) (1970–4). *Dictionary of scientific biography*. 15 vols. New York, Charles Scribner's Sons.

Gingerich, O. (1980). Was Ptolemy a fraud? *Q. J. R. Astr. Soc.*, 21, 253–66.

—— (1981). Ptolemy revisited: a reply to R. R. Newton. *Q. J. R. Astr. Soc.*, 22, 40–4.

Goldstein, B. R. (1964). On the theory of trepidation. *Centaurus*, 10, 232–47.

Gordon, C. H. (1967). *The ancient near east.* 3rd rev. edn. New York, W. W. Norton.

Gössmann, F. (1950). *Planetarium Babylonicum.* Band 2, Teil IV of A. Deimal, *Sumerisches Lexicon.* Rome, Pap stl. Bibelinstituts.

Gunther, R. T. G. (1976). *Astrolabes of the world.* 2 vols. 2nd edn, Vol. 1. London, The Holland Press. Originally published by Oxford University Press, 1932.

Hartner, W. (1965). The earliest history of the constellations and the motif of the lion-bull combat. *J. Near Eastern Stud.,* 24, 1–16.

—— (1977). The role of observations in ancient and modern astronomy. *J. Hist. Astr.,* 8, 1–11.

Hawkins, G. (1965). *Stonehenge decoded.* New York, Doubleday.

Heath, T. (1913). *Aristarchos of Samos, the ancient Copernicus.* Oxford, Clarendon Press.

Herity, M. (1974). *Irish passage graves.* Dublin, Irish University Press.

Hipparchos. *Aratou kai Eudoxou phainomenon exegegeos. (The commentary on the phenomena of Aratos and Eudoxos).* Edited and translated into German by C. Manitius (1894). Leipzig, Teubner.

Hodson, F. R. (ed.) (1974). *The place of astronomy in the ancient world.* London, Oxford University Press.

Hooke, S. H. (1963). *Babylonian and Assyrian religion.* Norman, University of Oklahoma Press.

Hoyle, F. (1972). *From Stonehenge to modern cosmology.* San Francisco, W. H. Freeman.

Kendall, D. G. (1974). Hunting quanta. In Hodson 1974.

Kennedy, E. S. (1966). Late medieval planetary theory. *Isis,* 57, 365–78.

Kennedy, E. S. and Roberts, V. (1959). The planetary theory of al-Shatir. *Isis,* 50, 227–35.

Kopal, Z. (1973). In Bienkowska 1973.

Knobel, E. G. (1917). *Ulugh Beg's catalogue of stars.* Carnegie Institution of Washington, publ. no. 251.

Kramer, S. N. (1959). *History begins at Sumer.* New York, Doubleday.

Kren, C. (1971). The rolling device of Nasir al-Din al-Tusi in the *De Spera* of Nicole Oresme. *Isis,* 62, 490–8.

Krupp, E. C. (ed.) (1979). *In search of ancient astronomers.* New York, McGraw-Hill.

Kunitzsch, P. (1983). How we got our 'Arabic' star names. *Sky Telesc.,* 65, 20–2.

Landes, D. S. (1983). *Revolution in time.* Cambridge, Mass., Harvard University Press.

Langdon, S. (1921). The early chronology of Sumer and Egypt and the similarities in their culture. *J. Egypt. Archaeol.,* 7, 133–53.

—— (1923). *The Babylonian epic of creation.* Oxford, Clarendon Press.

Langdon, S. and Fotheringham, J. K. with Schoch, C. (1928). *The Venus tablets of Ammizaduga*. London, Oxford University Press.

Lewis, D. (1972). *We, the navigators*. Canberra, Australian National University Press.

—— (1974). Voyaging stars: aspects of Polynesian and Micronesian astronomy. In Hodson 1974.

Lloyd, G. E. R. (1970). *Early Greek science: Thales to Aristotle*. London, Chatto and Windus.

—— (1973). *Greek science after Aristotle*. London, Chatto and Windus.

Luckenbill, D. D. (1923). Akkadian origins. *Amer. J. Semit. Lang. Lit.*, 40, 1–13.

Mackie, E. W. (1974). Archaeological tests on supposed prehistoric astronomical sites in Scotland. In Hodson 1974.

—— (1977). *The megalith builders*. London, Phaidon Press.

Nasr, S. H. (1968). *Science and civilization in Islam*. Cambridge, Mass., Harvard University Press.

—— (1976). *Islamic science: an illustrated study*. World of Islam Festival Publishing Company.

Needham, J. (1959). *Science and civilization in China*, Vol. 3. Cambridge University Press.

Neugebauer, O. (1947). Studies in ancient astronomy. VIII, The water clock in Babylonian astronomy. *Isis*, 37, 37–43.

—— (1949). The early history of the astrolabe. *Isis*, 40, 240–56.

—— (1951). The Babylonian method for the calculation of the last visibility of Mercury. *Proc. Am. Phil. Soc.*, 95, 110–16.

—— (1969). *The exact sciences of antiquity*. 2nd edn, corrected. New York, Dover.

—— (1975). *A history of ancient mathematical astronomy*, in 3 parts. Berlin, Heidelberg and New York, Springer-Verlag.

Newton, R. R. (1977). *The crime of Claudius Ptolemy*. Baltimore and London, Johns Hopkins University Press.

—— (1980). Comments on 'was Ptolemy a fraud?' by Owen Gingerich. *Q. J. R. Astr. Soc.*, 21, 388–99.

Nilsson, N. M. P. (1920). *Primitive time-reckoning*. Lund, Gleerup.

North, J. D. (ed.) (1971). *Richard of Wallingford* (the Latin texts, English translations and commentaries). Oxford, Clarendon Press.

Norton's star atlas. (1973). 16th edn, ed. G. E. Satterthwaite. Edinburgh, Gall and Inglis.

O'Leary, D. L. (1949). *How Greek science passed to the Arabs*. London, Routledge and Kegan Paul.

O'Neil, W. M. (1975). *Time and the calendars*. Sydney University Press.

—— (1978). The star catalogues of Ptolemy and Copernicus. *Search*, 9, 230–3.

—— (1980). The origins of Ptolemy's star catalogue. *Search*, 11, 112–14.

Oppenheim, A. L. (1964). *Ancient Mesopotamia*. University of Chicago Press.

—— (1974). Man and nature in Mesopotamian civilization. In *Dictionary of scientific biography*, ed. C. C. Gillespie. Vol. XV. New York, Charles Scribner's Sons.

Oppolzer, T. von [1887] (1962). *Canon of eclipses*. English translation by O. Gingerich of the original German text. New York, Dover.

Pannekoek, A. (1961). *A history of astronomy*. English translation from the Dutch. New York, Interscience.

Parker, R. A. (1974). Ancient Egyptian astronomy. In Hodson 1974.

Patrick, J. (1981). A reassessment of the solstitial observatories at Kintraw and Ballochroy. In Ruggles and Whittle 1981.

Peters, C. H. F. and Knobel, E. B. (1915). *Ptolemy's catalogue of stars*. Carnegie Institute of Washington, publ. no. 86.

Petersen, V. M. (1966). A comment on a comment by Manitius. *Centaurus*, 11, 306–9.

Petersen, V. M. and Schmidt, O. (1968). The determination of the apogee of the orbit of the Sun according to Hipparchos and Ptolemy. *Centaurus*, 12, 73–96.

Pingree, D. E. (1963). Astronomy and astrology in India and Iran. *Isis*, 54, 229–46.

—— (1968). *The thousands of Abu Ma'shar*. London, Warburg Institute.

—— (1974a). Astrology. *Encyclopaedia Britannica Macropaedia*. Vol. 2.

—— (1974b). History of mathematical astronomy in India. In *Dictionary of scientific biography*, ed. C. C. Gillespie. Vol. XV. New York, Charles Scribner's Sons.

Pliny, the Elder. *Natural history*. English translation by P. Holland (1942). Sussex, Centaur Press.

Price, D. J. de Solla (1955). *The equatorie of the planetis*. Cambridge University Press.

—— (1957). Precision instruments to 1500. In C. Singer *et al.* (eds) *A history of technology*. Vol. 3. Oxford, Clarendon Press.

—— (1974). Gears from the Greeks. The Antikythera mechanism – a calendar computer from *ca.* 80 BC. *Trans. Am. Phil. Soc.*, new series, 64, 7.

Ptolemaios, K. (*c.* AD 150). *Syntaxis mathematike*. English translations using Arabian title *Almagest* by (a) R. C. Taliaferro (1948), Chicago, Encyclopaedia Britannica, and (b) G. J. Toomer (1984), London, Duckworth.

—— (*c.* AD 150). *Tetrabiblos*. Edited and translated into English by F. E. Robbins (1940). London, Heinemann.

Rawlins, D. (1982). Eratosthenes' geodesy unravelled. *Isis*, 73, 259–65.

Reiner, E. (1975). *Babylonian planetary omens.* Malibu, Calif., Undena Publications.

Renou, L. and Filliozat, J. (1953). *L'Indique classique: manuel des Estudes Indiennes.* Vol. 2. Paris, Imprimerie Nationale.

Roberts, V. (1957). The solar and lunar theory of Ibn ash-Shatir: a pre-Copernican model. *Isis,* 48, 428–32.

—— (1966). The planetary theory of Ibn al-Shatir: latitudes of the planets. *Isis,* 57, 208–19.

Rome, A. (ed.) (1931). *Commentaires de Pappus et de Theon d'Alexandre sur l'Almageste.* 3 vols. Rome, Biblioteca Apostolica Vaticana.

Rosen, E. (1959). *Three Copernican treatises.* Including an English translation of *Commentariolus.* New York, Dover.

—— (1978). Translation of Copernicus, *De revolutionibus.* London, Macmillan.

Rudnicki, J. (1943). *Nicholas Copernicus.* English translation from Polish by B. W. A. Massey. London, Copernicus Quatercentenary Celebration Committee.

Ruggles, C. L. N. and Whittle, A. W. R. (eds) (1981). *Astronomy and society in Britain during the period 4000–1500* BC. Oxford, British Archaeological Reports.

Sachs, A. (1948). A classification of the Babylonian astronomical tablets of the Seleucid period. *J. Cuneiform Stud.,* 2, 211–90.

—— (1952a). Babylonian horoscopes. *J. Cuneiform Stud.,* 6, 49–75.

—— (1952b). Sirius dates in Babylonian astronomical texts of the Seleucid period. *J. Cuneiform Stud.,* 6, 105–11.

—— (1952c). A later Babylonian star catalogue. *J. Cuneiform Stud.,* 6, 146–50.

—— (1974). Babylonian observational astronomy. In Hodson 1974.

Sewell, R. (1924). *The Siddhantas and the Indian calendar.* Calcutta, Government of India Central Publication Branch.

Shulman, S. (1976). *The encyclopaedia of astrology.* London, Hamlyn.

Solinus, C. J. *Collectanea rerum memorabilium.* Translated into English by A. Golding (1955), *The excellent and pleasant works.* Gainesville, Florida, Scholars' Facsimiles and Reprints.

Stahl, W. H. (1945). The Greek heliocentric theory and its abandonment. *Trans. Am. Philol. Ass.,* 76, 321–32.

Taliaferro, R. C. (1935). Translation of Ptolemy, *Syntaxis mathematike.* Chicago, Encyclopaedia Britannica.

Thom, A. (1966). Megaliths and mathematics. *Antiquity,* 40, 121–8.

—— (1967). *Megalithic sites in Britain.* Oxford, Clarendon Press.

—— (1971). *Megalithic lunar observatories.* Oxford, Clarendon Press.

Thom, A. *et al.* (1974). Stonehenge. *J. Hist. Astr.,* 5, 71–90.

—— (1975). Stonehenge as a possible lunar observatory. *J. Hist. Astr.,* 6, 19–30.

Thom, A. and Thom, A. S. (1978). *Megalithic remains in Britain and Brittany.* Oxford, Clarendon Press.

Toomer, G. J. (1984). *Ptolemy's Almagest.* An English translation of *Syntaxis mathematike.* London, Duckworth.

Tuckerman, B. (1962 and 1964). *Planetary lunar and solar positions.* Vol. 1, 601 BC to AD 1. Vol. 2, AD 2 to AD 1649. Philadelphia, American Philosophical Society.

van der Waerden, B. L. (1949). Babylonian astronomy. II. The thirty-six stars. *J. Near Eastern Stud.,* 8, 6–26.

—— (1951). Babylonian astronomy. III. The earliest astronomical computations. *J. Near Eastern Stud.,* 10, 20–34.

Vogt, N. (1925). Versuch einer Widerherstellung von Hipparchs Fixsternverzeichnis. *Astr. Nachr.,* Bd. 221, cols. 17–54.

Wallis, C. G. (1939). Translation of Copernicus, *De revolutionibus.* Chicago, Encyclopaedia Britannica.

Weir, J. D. (1972). *The Venus tablets of Ammizaduga.* Istanbul and Leiden, L'Institute historique et archéologique de Nederlands.

Wood, J. E. (1978). *Sun, moon and standing stones.* Oxford University Press.

INDEX

Ahmad ibn Yunus, 121–2, 125, 126
al-Balki, 120
al-Battani (Albategnius), 120–1, 125, 126, 148, 172
Albertus Magnus, 148
al-Besri, 122
al-Biruni, 122
al-Bitruji, 124, 128, 147, 148
al-Buzjani, 121
al-Farghani, 120
al-Haytham, 122
al-Khayyami (Omar Khayyam), 122–3
al-Khwarizmi, 120
Allen, R. H., 1899, 162
al-Mansur, 107
al-Sabi, 120
al-Shatir, 119, 124, 128–30, 169–70
al-Shirazi, 119, 123, 124, 128
al-Sufi, 121
al-Tusi, 119, 124, 128, 130, 169
al-Zargali, 122, 125, 126, 127
Alleged astronomical alignments of megalithic monuments, 186–92
Alleged megalithic observatories, 181–95
Alleged uses of Stonehenge as an eclipse predictor, 192–4

Anomolous apparent motions of the 'planets', 3–8
Antikythera computer, 144–5
Anu's, Enlil's and Ea's ways, 19, 20
Apparent motions of celestial bodies, fixed stars and 'planets', 1–8
Arabian astronomy, 117–32
Arabian observatories, 130–2
Archimedes' (Jacob's) staff, 140
Aristarchos, 57–8, 60
Armillae, 141–3
Assurbanipal, 21
Astrolabe (planispheric), 143–4
Astrology, 12, 109–16
Atkinson, R. J. C., 1956, 184; 1966a, b, 191
Aveni, F. A., 1977, 192

Babylonian and Assyrian Akkadian speakers, 16–17, 19
Babylonian contributions to Greek astronomy, 50–1
Babylonian 'diaries', 26, 35–6
Babylonian ephemerides, 41–50
Babylonian 'goal texts', 27–9, 40
Babylonian mathematics, 37–9
Babylonian 'step functions', 44–6, 106
Babylonian 'zigzag functions', 46–7, 49–50, 106

Babylonian zodiacal signs, 40–1
Ballinaby, 189 ˇ
Ballochroy, 187
Boll, F., 1901, 73
Brahe, Tycho, *see* Tycho Brahe
Burl, A., 1976, 184, 186

Calcagnini, Celio, 151
Campanus of Novara, 149
Carnac, 190
Celestial equator, 2, 5
Chinese astronomy, 176–80; differences from the West, 176; independent discoveries, 177–8; influences from the West, 179–80; similarities with the West, 178–80
Circumpolar stars, 1, 179
Colton, R. and Martin, R. L., 1967, 1969, 193–4
Commentaries on Ptolemy's *Syntaxis mathematike* by Pappos, Theon and Proklos, 99–100; by Arabian astronomers, 130
Commentary on *The Phenomena of Aratos and Eudoxos* by Hipparchos, 68
Comparison of Babylonian and Greek astronomy, 50–1, 52–3, 54–5
Concentric planetary spheres, 61–6, 124
Conditions affecting the directions of the rising and setting of celestial bodies, 2–6, 182–4
Constellation and star names and their origins, 17–18, 153–63
Copernicus, 164–75; assumption of the rotation and revolution of the Earth, 164; as the last great ancient astronomer, 164; elimination of equants and crank-mechanism, 169–70; explanation of anomalous apparent 'planetary' motion, 165–7; increase over the number of Ptolemy's circles, 173; parallels with al-Shatir's planetary theories, 119, 169–70; unification of the planetary theories, 167
Crombie, A. C., 1953, 148

Delambre, J-B. J., 95
de Saussure, L., 1919–20, 105
Dicks, D. R., 1970, 53
Diopter, 140
Distinction between tropical and sidereal years, 70–2
Divination by the stars (astrology), 12, 109–16
Dreyer, J. L. E., 1905, 53, 60, 61, 67, 83, 95, 1917–18, 75

Early Christian Fathers, cosmological views of, 101–2
Early Greek observational astronomy, 55–60
Early Latin encyclopaedia, 100
Early uses of celestial phenomena, 8–12
Easter, problems of computing the date of, 102–3, 149–50
Eclipse cycles: Babylonian, 25, 29–34; Hipparchan, 69
Ecliptic, 3–5
Equatoria, 149
Equinoxes and solstices, 3, 56–7, 69–70, 125–6, 187–8
Eratosthenes, 58–9, 139–40, 146, 172
Eudoxos, 55, 61–5, 66, 101
Euktemon, 56

Fotheringham, J. K., 1928, 23
Fracastoro, Girolamo, 151

Galileo, 171, 174
·Geminos, 78–80
Gillespie, C. C., 1970–4, 53, 67, 119
Gingerich, O., 1980, 1981, 95
Gordon, C. H., 1967, 16
Gössmann, F., 1950, 19, 157, 158–9, 162
Greek prescientific astronomical speculation, 53–5
Greek units of angular measurement, 66, 130
Grosseteste, R., 147–8
Gundel, W., 1936, 73, 74
Gunther, R. T. G., 1976, 144

Halley, E., 33
Hartner, W., 1965, 10, 17; 1977, 97
Hawkins, G., 1965, 191, 194
Heraklides's theory, 100–1
Herity, M., 1974, 186
Hesiod, 52
Hipparchos, 55, 56, 57, 67–79, 172
Hooke, S. H., 1963, 16, 110
Hoyle, F., 1972, 193, 194

Indian astronomy in the first millenium AD 104–7

Jabir ibn Aflah al-Ishbili, 123, 128
Jai Singh, 132, 195
Jacob's (Archimedes's) staff, 140, 151
Johannes de Sacrobosco, 148

Kallippic cycle, 80, 145

Kallippos, 55, 56–7, 65, 66, 68, 69–70
Kendall, D. G., 1974, 185
Kennedy, E. S., 1966, 119, 128; 1959, 130
Kepler, J., 116, 164, 174–5, 178; his amendments to Copernican planetary theory, 174–5
Kintraw, 187
Klaudios Ptolemaios, *see* Ptolemy
Kleostratos, 187
Knobel, E. B., 1917, 128
Kopal, Z., 1973, 164
Kramer, S. N., 1959, 16
Krupp, E. C., 1979, 191, 192
Kunitzsch, P., 1983, 161

Landes, D. S., 1983, 138
Langdon, S., 1923, 18, 23
Lewis, D., 1972, 1975, 11
Lloyd, G. E. R., 1970, 1973, 53, 67, 83

Mackie, E. W., 1977, 185
Martianus Capella, 100–1
Masha'allah, 115, 120
Medieval translations of Arabian texts into Latin, 147
Megalithic monuments in Britain and Brittany as positive observatories, 185–95
Mesopotamian star names, 19
Metonic cycle, 55, 80, 103, 105, 145
Michelson, A. A. and Hanbury Brown, R., 174
Milesian philosophers, Thales, Anaximander and Anaximenes, 53–4

Nakshatra stars, 104–5

Nasr, S. H., 1968, 1976, 119
Navigation and orientation by the stars, 11–12, 114–15
Needham, J., 1959, 176, 178
Neugebauer, O., 1969, 8, 13, 37, 39, 53, 83; 1975, 13, 53, 67, 72, 73, 74
New Grange passage tombs, 186
Newton, R. R., 1977, 75, 95, 126; 1980, 95
Nicholas of Cusa, 150
Nilsson, N. M. P., 1920, 8

Obliquity of the ecliptic, 3, 39, 95, 126–7, 172
O'Leary, D. L., 1949, 104, 107, 118
O'Neil, W. M., 1975, 8, 11, 56
Oppenheim, A. L., 1964, 16
Oppolzer, T. von, 1887, 97
Oresme, Nicole, 149, 169

Pannakoek, A., 1961, 53, 67, 83, 95
Parker, R. A., 1974, 8
Parmenides, 54
Patrick, J., 1981, 187
Peters, C. H. F. and Knobel, E. B., 1915, 128
Petersen, V. M., 1966, 70, 96
Peuerbach, Georg, 150–1
Pingree, D. E., 1963, 104, 110; 1974a, 12, 110; 1974b, 104
Planetary theories in terms of deferents and epicycles: Apolloonios, 76; Hipparchos, 76–8; Ptolemy, 85–93; Arabian, 128–9; Copernicus, 167–71
Plato, 55, 61
Pliny's *Natural History*, 100
Precession of the equinoxes, 71–2, 95–6, 172
Prescientific, proto-scientific and scientific astronomy, 12–14

Price, D. J. de Solla, 1955, 115; 1957, 1974, 145
Ptolemy, 55, 82–98, 107, 133–4, 72: arguments for a spherical, geocentric universe; as a questionable observer, 94–7; astrological contentions, 97–8, 111–13, 114; concept of the equant, 89–91; *Geographical Directory*, 97; *Handy Tables*, 98, 99; lunar theory, 85–7; *The Planetary Hypotheses*, 98, 118; solar theory, 85; stellar magnitudes, 133–4; *Syntaxis mathematike (Almagest)*, 81, 82–98, 108; *Tetrabiblos*, 97, 118, 120, 121, 122, 123; theory of Mercury, 91–3; the theories of Venus, of Mars, of Jupiter, and of Saturn, 87–91
Pythagoras, 54

Quadrant and sextant, 140

Rawlins, D., 1982, 66
Regiomontanus, 144, 150–1, 179
Reiner, A., 1975, 23
Relative distances of the 'planets', 57–8, 59–60
Richard of Wallingford, 149
Roberts, V., 1957, 119, 130; 1959, 130
Rome, A., 1931, 99
Rosen, E., 1959, 166

Sachs, A., 1948, 26, 35; 1952a, 12, 110
Schiaparelli, G. V., 61
Seleukos, 60, 164
Selinus's *Collection of Memorable Facts*, 100

Shadow-clocks, 135–6

Solar apogee: Ahmad ibn Yunus, 125, al-Battani, 125; al-Zargali, 125; Hipparchos, 76–7; Ptolemy, 96

Stahl, W. H., 1945, 100

Star catalogues: al-Sufi's, 121; Copernicus's, 174; Hipparchos's and Ptolemy's, 72–6; Ulugh Beg's, 127–8

Stellar magnitudes, 133–4

Stonehenge, 185, 190–4

Sumerians, 15–17, 190–4

Taliaferro, R. C., 1935, 82

Temple Wood, 190

Thabit ibn Qurra, 120

The Bull-of-Heaven (Taurus), 17–18

The thirty-six Mesopotamian reference stars, 17–18, 25–6

Thom, A., 1967, 1971, 1978, 184, 185, 188, 189; 1974, 191, 194; 1975, 191

Three stars a month in Mesopotamia, 18–19

Time-keeping devices, 134–8

Time-reckoning, 8–11, 179

Toledan tables, 122, 127, 149

Toomer, G. J., 1984, 82

Translations of Greek texts into Syriac by dissident Christians, 103–4

Trepidation of the equinoxes, 120

Triquetrum, 140–1

Tuckerman, B., 1962, 1964, 97, 103

Tycho Brahe, 144, 172, 173–4, 175, 178, 179; his objections to the Copernican planetary theory, 173–4

Types of megalithic monuments in Britain and Brittany, 184–6

Ulugh Beg, 123, 125, 126, 127–8, 131, 144, 172, 179

van der Waerden, B. L., 1972, 18

Venus tablets of Ammisaduga, 21–4, 181

Vogt, N., 1925, 73, 74

Water-clocks, 136–7, 179

Weir, J. D., 1972, 24

Wood, J. E., 1978, 184

Zodiacal constellations in Mesopotamia, 17–18

Zodiacal signs, 25, 41–2